CHAPTER 1

DIRECT CURRENT GENERATORS

LEARNING OBJECTIVES

Upon completion of the chapter you will be able to:

1. State the principle by which generators convert mechanical energy to electrical energy.

2. State the rule to be applied when you determine the direction of induced emf in a coil.

3. State the purpose of slip rings.

4. State the reason why no emf is induced in a rotating coil as it passes through a neutral plane.

5. State what component causes a generator to produce direct current rather than alternating current.

6. Identify the point at which the brush contact should change from one commutator segment to the next.

7. State how field strength can be varied in a dc generator.

8. Describe the cause of sparking between brushes and commutator.

9. State what is meant by "armature reaction."

10. State the purpose of interpoles.

11. Explain the effect of motor reaction in a dc generator.

12. Explain the causes of armature losses.

13. List the types of armatures used in dc generators.

14. State the three classifications of dc generators.

15. State the term that applies to voltage variation from no-load to full-load conditions and how it is expressed as a percentage.

16. State the term that describes the use of two or more generators to supply a common load.

17. State the purpose of a dc generator that has been modified to function as an amplidyne.

INTRODUCTION

A generator is a machine that converts mechanical energy into electrical energy by using the principle of magnetic induction. This principle is explained as follows:

Whenever a conductor is moved within a magnetic field in such a way that the conductor cuts across magnetic lines of flux, voltage is generated in the conductor.

The AMOUNT of voltage generated depends on (1) the strength of the magnetic field, (2) the angle at which the conductor cuts the magnetic field, (3) the speed at which the conductor is moved, and (4) the length of the conductor within the magnetic field.

The POLARITY of the voltage depends on the direction of the magnetic lines of flux and the direction of movement of the conductor. To determine the direction of current in a given situation, the LEFT-HAND RULE FOR GENERATORS is used. This rule is explained in the following manner.

Extend the thumb, forefinger, and middle finger of your left hand at right angles to one another, as shown in figure 1-1. Point your thumb in the direction the conductor is being moved. Point your forefinger in the direction of magnetic flux (from north to south). Your middle finger will then point in the direction of current flow in an external circuit to which the voltage is applied.

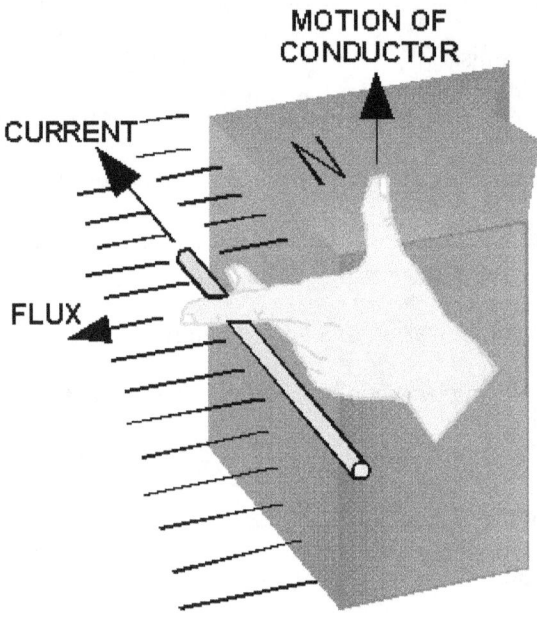

Figure 1-1.—Left-hand rule for generators.

THE ELEMENTARY GENERATOR

The simplest elementary generator that can be built is an ac generator. Basic generating principles are most easily explained through the use of the elementary ac generator. For this reason, the ac generator will be discussed first. The dc generator will be discussed later.

An elementary generator (fig. 1-2) consists of a wire loop placed so that it can be rotated in a stationary magnetic field. This will produce an induced emf in the loop. Sliding contacts (brushes) connect the loop to an external circuit load in order to pick up or use the induced emf.

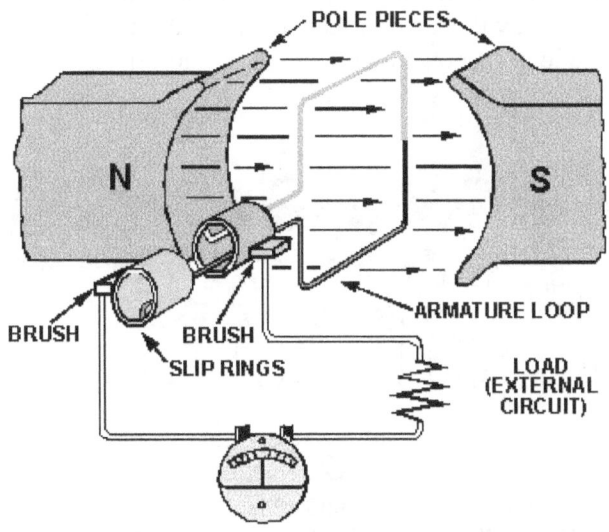

Figure 1-2.—The elementary generator.

The pole pieces (marked N and S) provide the magnetic field. The pole pieces are shaped and positioned as shown to concentrate the magnetic field as close as possible to the wire loop. The loop of wire that rotates through the field is called the ARMATURE. The ends of the armature loop are connected to rings called SLIP RINGS. They rotate with the armature. The brushes, usually made of carbon, with wires attached to them, ride against the rings. The generated voltage appears across these brushes.

The elementary generator produces a voltage in the following manner (fig. 1-3). The armature loop is rotated in a clockwise direction. The initial or starting point is shown at position A. (This will be considered the zero-degree position.) At 0° the armature loop is perpendicular to the magnetic field. The black and white conductors of the loop are moving parallel to the field. The instant the conductors are moving parallel to the magnetic field, they do not cut any lines of flux. Therefore, no emf is induced in the conductors, and the meter at position A indicates zero. This position is called the NEUTRAL PLANE. As the armature loop rotates from position A (0°) to position B (90°), the conductors cut through more and more lines of flux, at a continually increasing angle. At 90° they are cutting through a maximum number of lines of flux and at maximum angle. The result is that between 0° and 90°, the induced emf in the conductors builds up from zero to a maximum value. Observe that from 0° to 90°, the black conductor cuts DOWN through the field. At the same time the white conductor cuts UP through the field. The induced emfs in the conductors are series-adding. This means the resultant voltage across the brushes (the terminal voltage) is the sum of the two induced voltages. The meter at position B reads maximum value. As the armature loop continues rotating from 90° (position B) to 180° (position C), the conductors which were cutting through a maximum number of lines of flux at position B now cut through fewer lines. They are again moving parallel to the magnetic field at position C. They no longer cut through any lines of flux. As the armature rotates from 90° to 180°, the induced voltage will decrease to zero in the same manner that it increased during the rotation from 0° to 90°. The meter again reads zero. From 0° to 180° the conductors of the armature loop have been moving in the same direction through the magnetic field. Therefore, the polarity of the induced voltage has remained the same. This is shown by points A through C on the graph. As the loop rotates beyond 180° (position C), through 270° (position D), and back to the initial or starting point (position A), the direction of the cutting action of the conductors through the magnetic field reverses. Now the black conductor cuts UP through the field while the white conductor cuts DOWN through the field. As a result, the polarity of the induced voltage reverses. Following the sequence shown by graph points C, D, and back to A, the voltage will be in the direction opposite to that shown from points A, B, and C. The terminal voltage will be the same as it was from A to C except that the polarity is reversed (as shown by the meter deflection at position D). The voltage output waveform for the complete revolution of the loop is shown on the graph in figure 1-3.

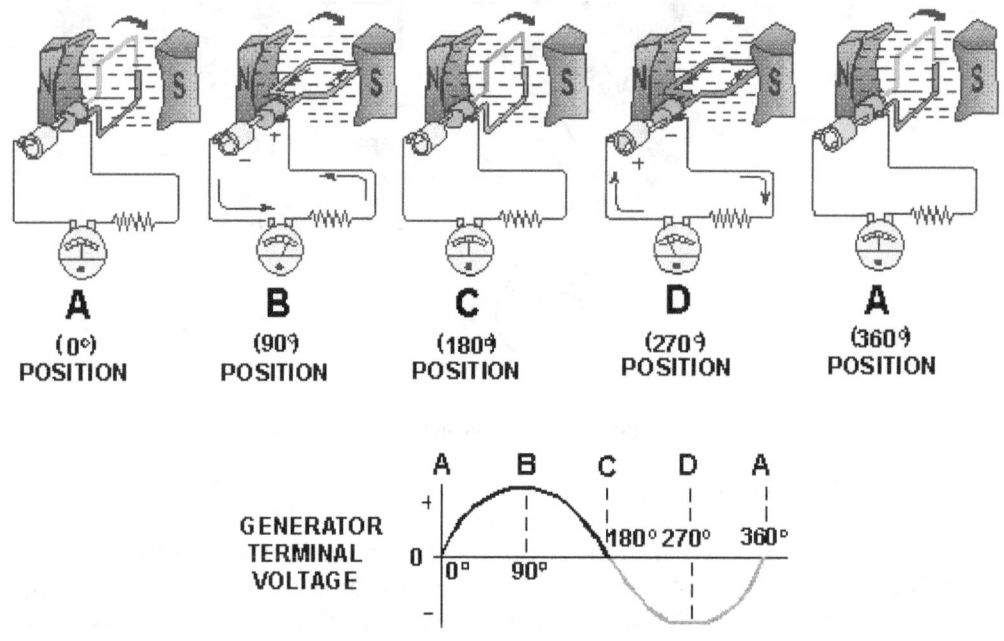

Figure 1-3.—Output voltage of an elementary generator during one revolution.

Q1. Generators convert mechanical motion to electrical energy using what principle?

Q2. What rule should you use to determine the direction of induced emf in a coil?

Q3. What is the purpose of the slip rings?

Q4. Why is no emf induced in a rotating coil when it passes through the neutral plane?

THE ELEMENTARY DC GENERATOR

A single-loop generator with each terminal connected to a segment of a two-segment metal ring is shown in figure 1-4. The two segments of the split metal ring are insulated from each other. This forms a simple COMMUTATOR. The commutator in a dc generator replaces the slip rings of the ac generator. This is the main difference in their construction. The commutator mechanically reverses the armature loop connections to the external circuit. This occurs at the same instant that the polarity of the voltage in the armature loop reverses. Through this process the commutator changes the generated ac voltage to a pulsating dc voltage as shown in the graph of figure 1-4. This action is known as commutation. Commutation is described in detail later in this chapter.

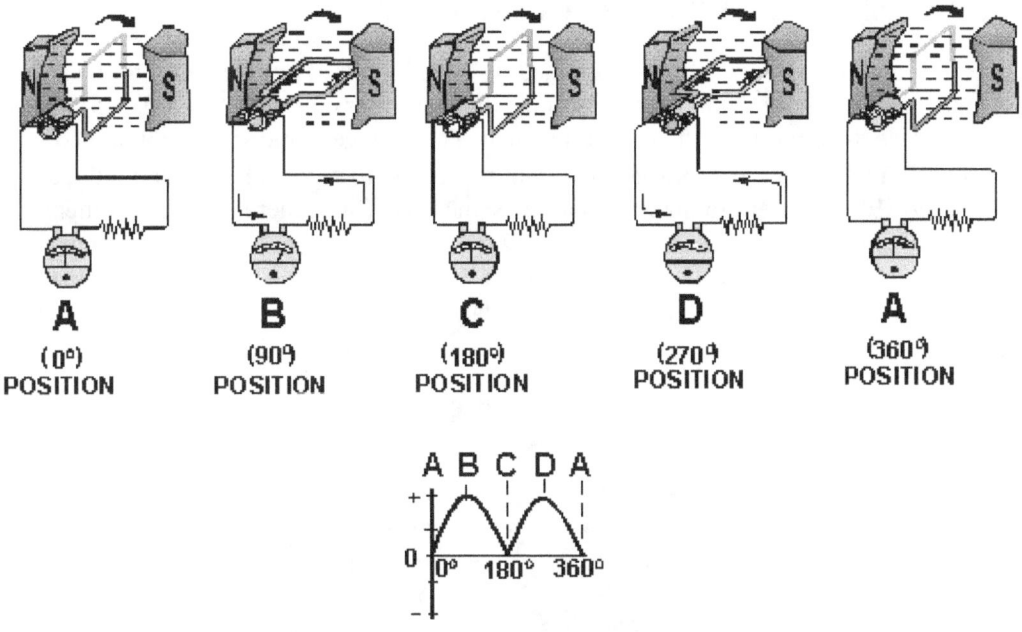

Figure 1-4.—Effects of commutation.

For the remainder of this discussion, refer to figure 1-4, parts A through D. This will help you in following the step-by-step description of the operation of a dc generator. When the armature loop rotates clockwise from position A to position B, a voltage is induced in the armature loop which causes a current in a direction that deflects the meter to the right. Current flows through loop, out of the negative brush, through the meter and the load, and back through the positive brush to the loop. Voltage reaches its maximum value at point B on the graph for reasons explained earlier. The generated voltage and the current fall to zero at position C. At this instant each brush makes contact with both segments of the commutator. As the armature loop rotates to position D, a voltage is again induced in the loop. In this case, however, the voltage is of opposite polarity.

The voltages induced in the two sides of the coil at position D are in the reverse direction to that of the voltages shown at position B. Note that the current is flowing from the black side to the white side in position B and from the white side to the black side in position D. However, because the segments of the commutator have rotated with the loop and are contacted by opposite brushes, the direction of current flow through the brushes and the meter remains the same as at position B. The voltage developed across the brushes is pulsating and unidirectional (in one direction only). It varies twice during each revolution between zero and maximum. This variation is called RIPPLE.

A pulsating voltage, such as that produced in the preceding description, is unsuitable for most applications. Therefore, in practical generators more armature loops (coils) and more commutator segments are used to produce an output voltage waveform with less ripple.

Q5. *What component causes a generator to produce dc voltage rather than ac voltage at its output terminals?*

Q6. *At what point should brush contact change from one commutator segment to the next?*

Q7. *An elementary, single coil, dc generator will have an output voltage with how many pulsations per revolution?*

EFFECTS OF ADDING ADDITIONAL COILS AND POLES

The effects of additional coils may be illustrated by the addition of a second coil to the armature. The commutator must now be divided into four parts since there are four coil ends (see fig. 1-5). The coil is rotated in a clockwise direction from the position shown. The voltage induced in the white coil, DECREASES FOR THE NEXT 90° of rotation (from maximum to zero). The voltage induced in the black coil INCREASES from zero to maximum at the same time. Since there are four segments in the commutator, a new segment passes each brush every 90° instead of every 180°. This allows the brush to switch from the white coil to the black coil at the instant the voltages in the two coils are equal. The brush remains in contact with the black coil as its induced voltage increases to maximum, level B in the graph. It then decreases to level A, 90° later. At this point, the brush will contact the white coil again.

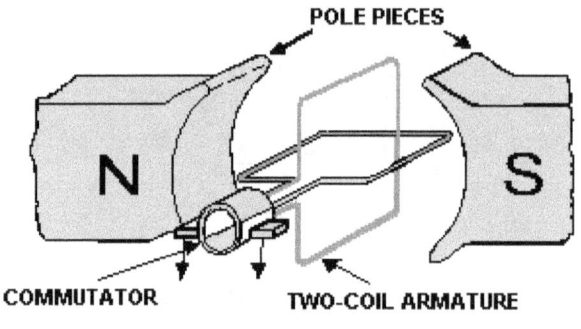

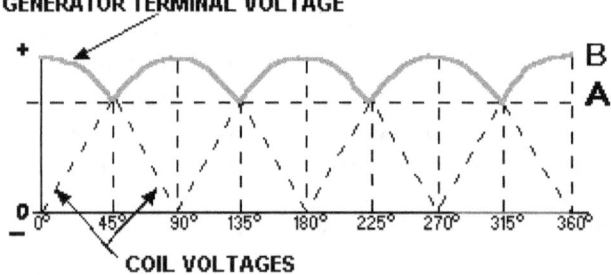

Figure 1-5.—Effects of additional coils.

The graph in figure 1-5 shows the ripple effect of the voltage when two armature coils are used. Since there are now four commutator segments in the commutator and only two brushes, the voltage cannot fall any lower than at point A. Therefore, the ripple is limited to the rise and fall between points A and B on the graph. By adding more armature coils, the ripple effect can be further reduced. Decreasing ripple in this way increases the effective voltage of the output.

NOTE: Effective voltage is the equivalent level of dc voltage, which will cause the same average current through a given resistance. By using additional armature coils, the voltage across the brushes is not allowed to fall to as low a level between peaks. Compare the graphs in figure 1-4 and 1-5. Notice that the ripple has been reduced. Practical generators use many armature coils. They also use more than one pair of magnetic poles. The additional magnetic poles have the same effect on ripple as did the additional armature coils. In addition, the increased number of poles provides a stronger magnetic field (greater number of flux lines). This, in turn, allows an increase in output voltage because the coils cut more lines of flux per revolution.

Q8. How many commutator segments are required in a two-coil generator?

ELECTROMAGNETIC POLES

Nearly all practical generators use electromagnetic poles instead of the permanent magnets used in our elementary generator. The electromagnetic field poles consist of coils of insulated copper wire wound on soft iron cores, as shown in figure 1-6. The main advantages of using electromagnetic poles are (1) increased field strength and (2) a means of controlling the strength of the fields. By varying the input voltage, the field strength is varied. By varying the field strength, the output voltage of the generator can be controlled.

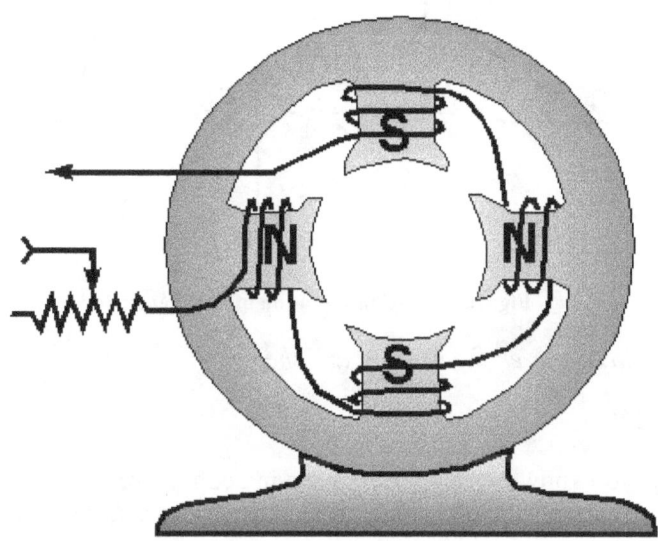

Figure 1-6.—Four-pole generator (without armature).

Q9. How can field strength be varied in a practical dc generator?

COMMUTATION

Commutation is the process by which a dc voltage output is taken from an armature that has an ac voltage induced in it. You should remember from our discussion of the elementary dc generator that the commutator mechanically reverses the armature loop connections to the external circuit. This occurs at the same instant that the voltage polarity in the armature loop reverses. A dc voltage is applied to the load because the output connections are reversed as each commutator segment passes under a brush. The segments are insulated from each other.

In figure 1-7, commutation occurs simultaneously in the two coils that are briefly short-circuited by the brushes. Coil B is short-circuited by the negative brush. Coil Y, the opposite coil, is short-circuited by the positive brush. The brushes are positioned on the commutator so that each coil is short-circuited as it moves through its own electrical neutral plane. As you have seen previously, there is no voltage generated in the coil at that time. Therefore, no sparking can occur between the commutator and the brush. Sparking between the brushes and the commutator is an indication of improper commutation. Improper brush placement is the main cause of improper commutation.

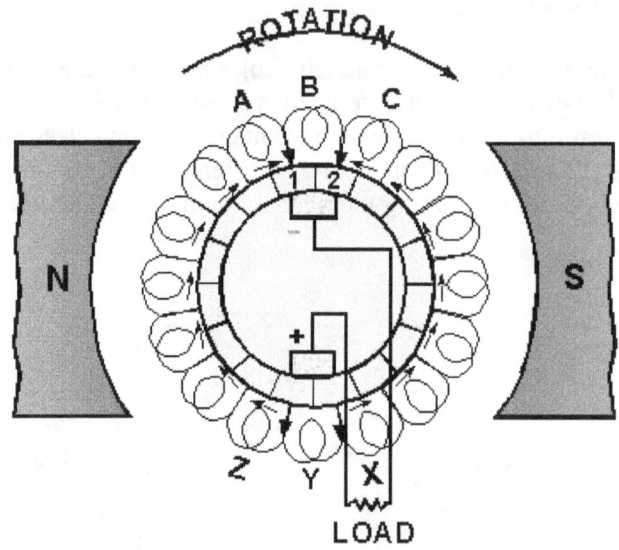

Figure 1-7.—Commutation of a dc generator.

Q10. What causes sparking between the brushes and the commutator?

ARMATURE REACTION

From previous study, you know that all current-carrying conductors produce magnetic fields. The magnetic field produced by current in the armature of a dc generator affects the flux pattern and distorts the main field. This distortion causes a shift in the neutral plane, which affects commutation. This change in the neutral plane and the reaction of the magnetic field is called ARMATURE REACTION.

You know that for proper commutation, the coil short-circuited by the brushes must be in the neutral plane. Consider the operation of a simple two-pole dc generator, shown in figure 1-8. View A of the figure shows the field poles and the main magnetic field. The armature is shown in a simplified view in views B and C with the cross section of its coil represented as little circles. The symbols within the circles represent arrows. The dot represents the point of the arrow coming toward you, and the cross represents the tail, or feathered end, going away from you. When the armature rotates clockwise, the sides of the coil to the left will have current flowing toward you, as indicated by the dot. The side of the coil to the right will have current flowing away from you, as indicated by the cross. The field generated around each side of the coil is shown in view B of figure 1-8. This field increases in strength for each wire in the armature coil, and sets up a magnetic field almost perpendicular to the main field.

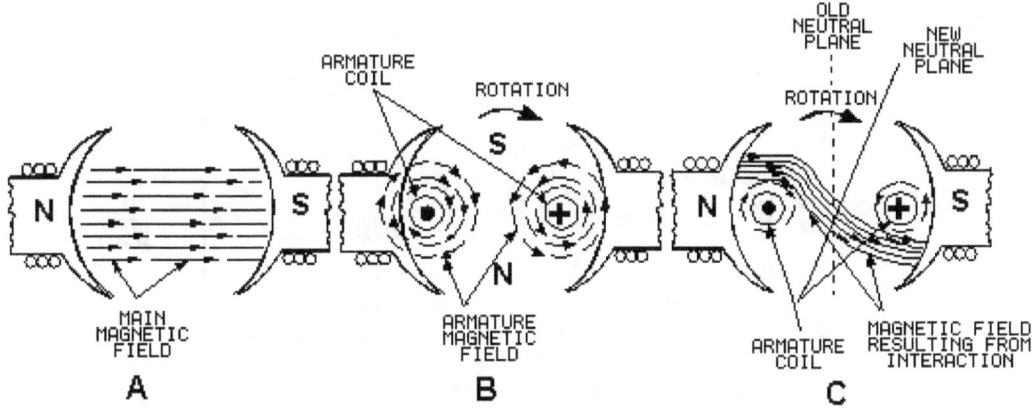

Figure 1-8.—Armature reaction.

Now you have two fields — the main field, view A, and the field around the armature coil, view B. View C of figure 1-8 shows how the armature field distorts the main field and how the neutral plane is shifted in the direction of rotation. If the brushes remain in the old neutral plane, they will be short-circuiting coils that have voltage induced in them. Consequently, there will be arcing between the brushes and commutator.

To prevent arcing, the brushes must be shifted to the new neutral plane.

Q11. What is armature reaction?

COMPENSATING WINDINGS AND INTERPOLES

Shifting the brushes to the advanced position (the new neutral plane) does not completely solve the problems of armature reaction. The effect of armature reaction varies with the load current. Therefore, each time the load current varies, the neutral plane shifts. This means the brush position must be changed each time the load current varies.

In small generators, the effects of armature reaction are reduced by actually mechanically shifting the position of the brushes. The practice of shifting the brush position for each current variation is not practiced except in small generators. In larger generators, other means are taken to eliminate armature reaction. COMPENSATING WINDINGS or INTERPOLES are used for this purpose (fig. 1-9). The compensating windings consist of a series of coils embedded in slots in the pole faces. These coils are connected in series with the armature. The series-connected compensating windings produce a magnetic field, which varies directly with armature current. Because the compensating windings are wound to produce a field that opposes the magnetic field of the armature, they tend to cancel the effects of the armature magnetic field. The neutral plane will remain stationary and in its original position for all values of armature current. Because of this, once the brushes have been set correctly, they do not have to be moved again.

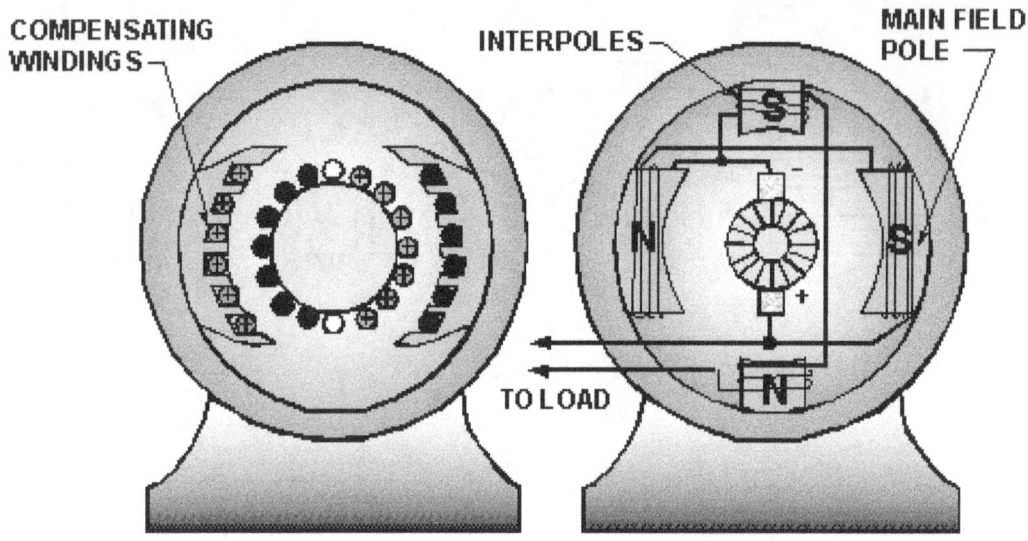

Figure 1-9.—Compensating windings and interpoles.

Another way to reduce the effects of armature reaction is to place small auxiliary poles called "interpoles" between the main field poles. The interpoles have a few turns of large wire and are connected in series with the armature. Interpoles are wound and placed so that each interpole has the same magnetic polarity as the main pole ahead of it, in the direction of rotation. The field generated by the interpoles produces the same effect as the compensating winding. This field, in effect, cancels the armature reaction for all values of load current by causing a shift in the neutral plane opposite to the shift caused by armature reaction. The amount of shift caused by the interpoles will equal the shift caused by armature reaction since both shifts are a result of armature current.

Q12. *What is the purpose of interpoles?*

MOTOR REACTION IN A GENERATOR

When a generator delivers current to a load, the armature current creates a magnetic force that opposes the rotation of the armature. This is called MOTOR REACTION. A single armature conductor is represented in figure 1-10, view A. When the conductor is stationary, no voltage is generated and no current flows. Therefore, no force acts on the conductor. When the conductor is moved downward (fig. 1-10, view B) and the circuit is completed through an external load, current flows through the conductor in the direction indicated. This sets up lines of flux around the conductor in a clockwise direction.

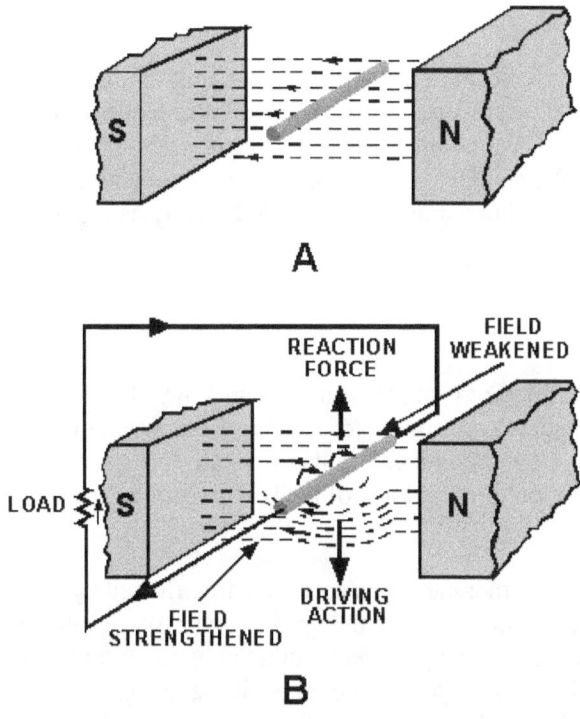

Figure 1-10.—Motor reaction in a generator.

The interaction between the conductor field and the main field of the generator weakens the field above the conductor and strengthens the field below the conductor. The main field consists of lines that now act like stretched rubber bands. Thus, an upward reaction force is produced that acts in opposition to the downward driving force applied to the armature conductor. If the current in the conductor increases, the reaction force increases. Therefore, more force must be applied to the conductor to keep it moving.

With no armature current, there is no magnetic (motor) reaction. Therefore, the force required to turn the armature is low. As the armature current increases, the reaction of each armature conductor against rotation increases. The actual force in a generator is multiplied by the number of conductors in the armature. The driving force required to maintain the generator armature speed must be increased to overcome the motor reaction. The force applied to turn the armature must overcome the motor reaction force in all dc generators. The device that provides the turning force applied to the armature is called the PRIME MOVER. The prime mover may be an electric motor, a gasoline engine, a steam turbine, or any other mechanical device that provides turning force.

Q13. What is the effect of motor reaction in a dc generator?

ARMATURE LOSSES

In dc generators, as in most electrical devices, certain forces act to decrease the efficiency. These forces, as they affect the armature, are considered as losses and may be defined as follows:

1. I^2R, or copper loss in the winding

2. Eddy current loss in the core

3. Hysteresis loss (a sort of magnetic friction)

Copper Losses

The power lost in the form of heat in the armature winding of a generator is known as COPPER LOSS. Heat is generated any time current flows in a conductor. Copper loss is an I^2R loss, which increases as current increases. The amount of heat generated is also proportional to the resistance of the conductor. The resistance of the conductor varies directly with its length and inversely with its cross-sectional area. Copper loss is minimized in armature windings by using large diameter wire.

Q14. What causes copper losses?

Eddy Current Losses

The core of a generator armature is made from soft iron, which is a conducting material with desirable magnetic characteristics. Any conductor will have currents induced in it when it is rotated in a magnetic field. These currents that are induced in the generator armature core are called EDDY CURRENTS. The power dissipated in the form of heat, as a result of the eddy currents, is considered a loss.

Eddy currents, just like any other electrical currents, are affected by the resistance of the material in which the currents flow. The resistance of any material is inversely proportional to its cross-sectional area. Figure 1-11, view A, shows the eddy currents induced in an armature core that is a solid piece of soft iron. Figure 1-11, view B, shows a soft iron core of the same size, but made up of several small pieces insulated from each other. This process is called lamination. The currents in each piece of the laminated core are considerably less than in the solid core because the resistance of the pieces is much higher. (Resistance is inversely proportional to cross-sectional area.) The currents in the individual pieces of the laminated core are so small that the sum of the individual currents is much less than the total of eddy currents in the solid iron core.

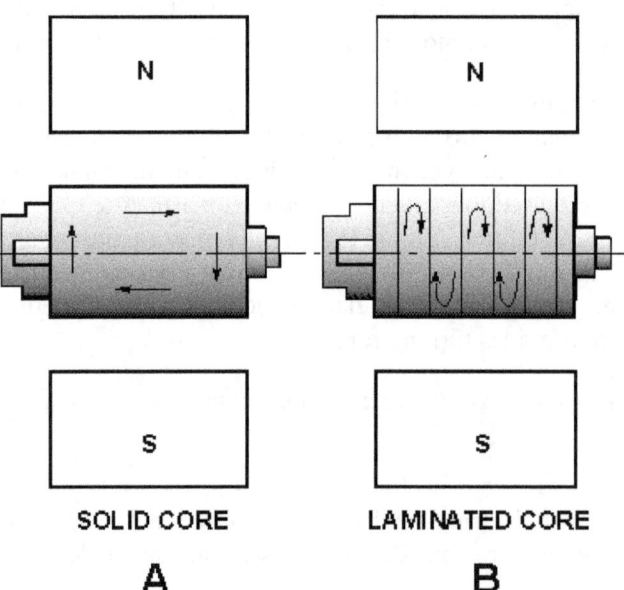

Figure 1-11.—Eddy currents in dc generator armature cores.

As you can see, eddy current losses are kept low when the core material is made up of many thin sheets of metal. Laminations in a small generator armature may be as thin as 1/64 inch. The laminations are insulated from each other by a thin coat of lacquer or, in some instances, simply by the oxidation of the surfaces. Oxidation is caused by contact with the air while the laminations are being annealed. The insulation value need not be high because the voltages induced are very small.

Most generators use armatures with laminated cores to reduce eddy current losses.

Q15. How can eddy current be reduced?

Hysteresis Losses

Hysteresis loss is a heat loss caused by the magnetic properties of the armature. When an armature core is in a magnetic field, the magnetic particles of the core tend to line up with the magnetic field. When the armature core is rotating, its magnetic field keeps changing direction. The continuous movement of the magnetic particles, as they try to align themselves with the magnetic field, produces molecular friction. This, in turn, produces heat. This heat is transmitted to the armature windings. The heat causes armature resistances to increase.

To compensate for hysteresis losses, heat-treated silicon steel laminations are used in most dc generator armatures. After the steel has been formed to the proper shape, the laminations are heated and allowed to cool. This annealing process reduces the hysteresis loss to a low value.

THE PRACTICAL DC GENERATOR

The actual construction and operation of a practical dc generator differs somewhat from our elementary generators. The differences are in the construction of the armature, the manner in which the armature is wound, and the method of developing the main field.

A generator that has only one or two armature loops has high ripple voltage. This results in too little current to be of any practical use. To increase the amount of current output, a number of loops of wire are used. These additional loops do away with most of the ripple. The loops of wire, called windings, are evenly spaced around the armature so that the distance between each winding is the same.

The commutator in a practical generator is also different. It has several segments instead of two or four, as in our elementary generators. The number of segments must equal the number of armature coils.

GRAMME-RING ARMATURE

The diagram of a GRAMME-RING armature is shown in figure 1-12, view A. Each coil is connected to two commutator segments as shown. One end of coil 1 goes to segment A, and the other end of coil 1 goes to segment B. One end of coil 2 goes to segment C, and the other end of coil 2 goes to segment B. The rest of the coils are connected in a like manner, in series, around the armature. To complete the series arrangement, coil 8 connects to segment A. Therefore, each coil is in series with every other coil.

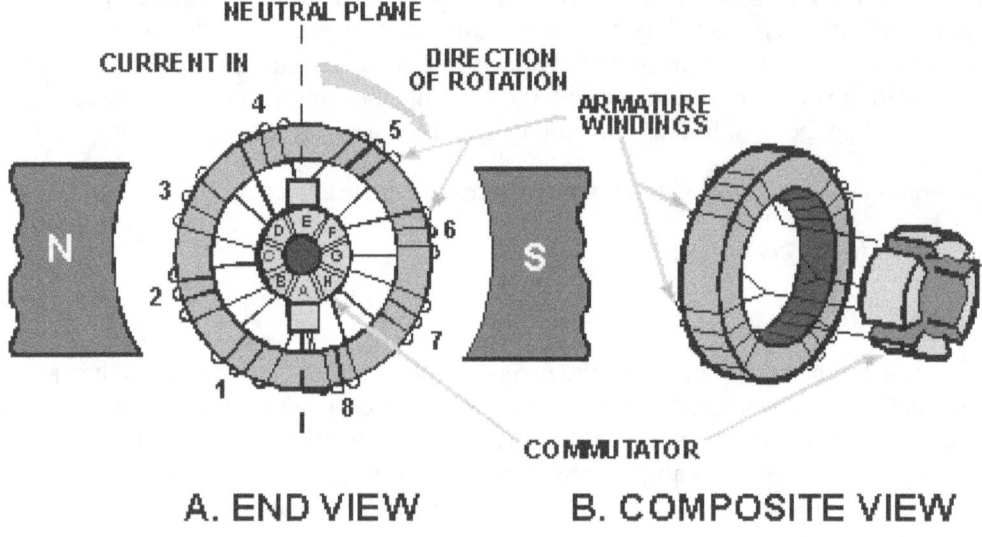

A. END VIEW   **B. COMPOSITE VIEW**

Figure 1-12.—Gramme-ring armature.

Figure 1-12, view B shows a composite view of a Gramme-ring armature. It illustrates more graphically the physical relationship of the coils and commutator locations.

The windings of a Gramme-ring armature are placed on an iron ring. A disadvantage of this arrangement is that the windings located on the inner side of the iron ring cut few lines of flux. Therefore, they have little, if any, voltage induced in them. For this reason, the Gramme-ring armature is not widely used.

DRUM-TYPE ARMATURE

A drum-type armature is shown in figure 1-13. The armature windings are placed in slots cut in a drum-shaped iron core. Each winding completely surrounds the core so that the entire length of the conductor cuts the main magnetic field. Therefore, the total voltage induced in the armature is greater than in the Gramme-ring. You can see that the drum-type armature is much more efficient than the Gramme-ring. This accounts for the almost universal use of the drum-type armature in modem dc generators.

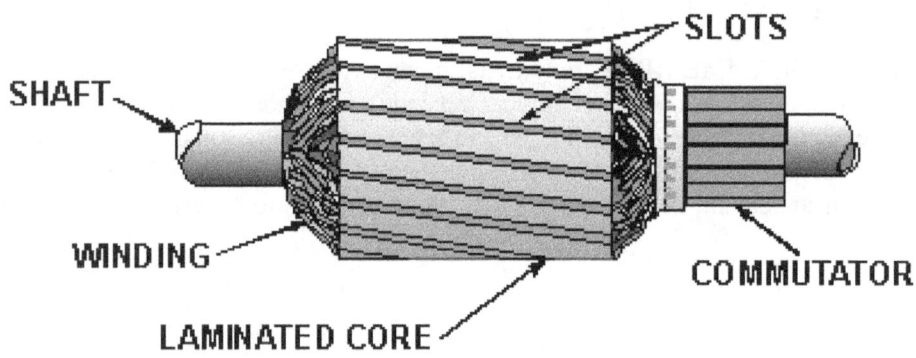

Figure 1-13.—Drum-type armature.

Drum-type armatures are wound with either of two types of windings — the LAP WINDING or the WAVE WINDING.

The lap winding is illustrated in figure 1-14, view A This type of winding is used in dc generators designed for high-current applications. The windings are connected to provide several parallel paths for current in the armature. For this reason, lap-wound armatures used in dc generators require several pairs of poles and brushes.

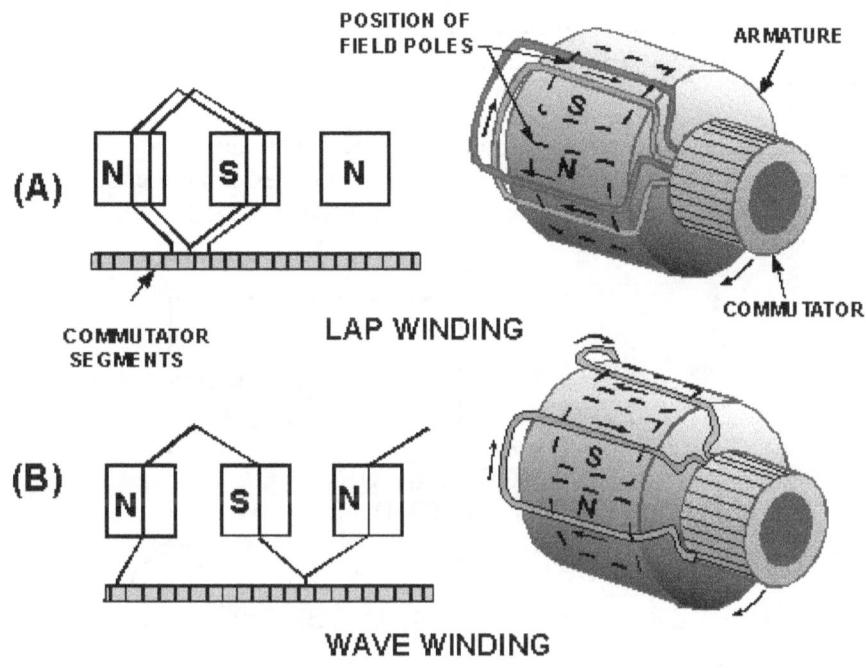

Figure 1-14.—Types of windings used on drum-type armatures.

Figure 1-14, view B, shows a wave winding on a drum-type armature. This type of winding is used in dc generators employed in high-voltage applications. Notice that the two ends of each coil are connected to commutator segments separated by the distance between poles. This configuration allows the series addition of the voltages in all the windings between brushes. This type of winding only requires one pair of brushes. In practice, a practical generator may have several pairs to improve commutation.

Q16. Why are drum-type armatures preferred over the Gramme-ring armature in modern dc generators?

Q17. Lap windings are used in generators designed for what type of application?

FIELD EXCITATION

When a dc voltage is applied to the field windings of a dc generator, current flows through the windings and sets up a steady magnetic field. This is called FIELD EXCITATION.

This excitation voltage can be produced by the generator itself or it can be supplied by an outside source, such as a battery. A generator that supplies its own field excitation is called a SELF-EXCITED GENERATOR. Self-excitation is possible only if the field pole pieces have retained a slight amount of permanent magnetism, called RESIDUAL MAGNETISM. When the generator starts rotating, the weak residual magnetism causes a small voltage to be generated in the armature. This small voltage applied to the field coils causes a small field current. Although small, this field current strengthens the magnetic field and allows the armature to generate a higher voltage. The higher voltage increases the field strength, and so on. This process continues until the output voltage reaches the rated output of the generator.

CLASSIFICATION OF GENERATORS

Self-excited generators are classed according to the type of field connection they use. There are three general types of field connections — SERIES-WOUND, SHUNT-WOUND (parallel), and COMPOUND-WOUND. Compound-wound generators are further classified as cumulative-compound and differential-compound. These last two classifications are not discussed in this chapter.

Series-Wound Generator

In the series-wound generator, shown in figure 1-15, the field windings are connected in series with the armature. Current that flows in the armature flows through the external circuit and through the field windings. The external circuit connected to the generator is called the load circuit.

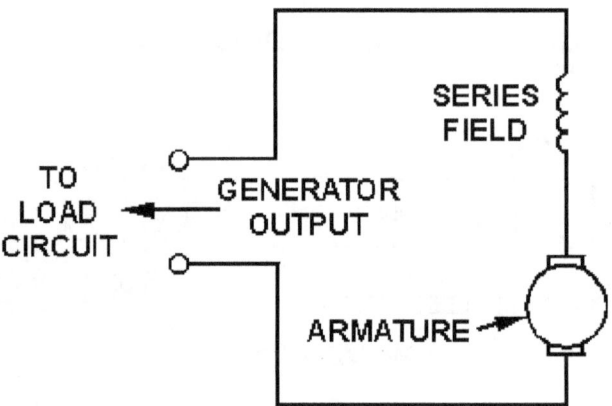

Figure 1-15.—Series-wound generator.

A series-wound generator uses very low resistance field coils, which consist of a few turns of large diameter wire.

The voltage output increases as the load circuit starts drawing more current. Under low-load current conditions, the current that flows in the load and through the generator is small. Since small current means that a small magnetic field is set up by the field poles, only a small voltage is induced in the armature. If the resistance of the load decreases, the load current increases. Under this condition, more current flows through the field. This increases the magnetic field and increases the output voltage. A series-wound dc generator has the characteristic that the output voltage varies with load current. This is undesirable in most applications. For this reason, this type of generator is rarely used in everyday practice.

The series-wound generator has provided an easy method to introduce you to the subject of self-excited generators.

Shunt-Wound Generators

In a shunt-wound generator, like the one shown in figure 1-16, the field coils consist of many turns of small wire. They are connected in parallel with the load. In other words, they are connected across the output voltage of the armature.

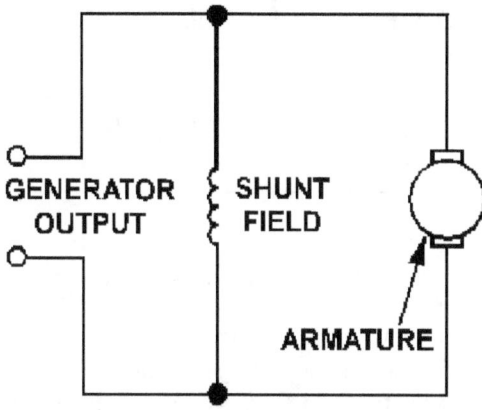

Figure 1-16.—Shunt-wound generator.

Current in the field windings of a shunt-wound generator is independent of the load current (currents in parallel branches are independent of each other). Since field current, and therefore field strength, is not affected by load current, the output voltage remains more nearly constant than does the output voltage of the series-wound generator.

In actual use, the output voltage in a dc shunt-wound generator varies inversely as load current varies. The output voltage decreases as load current increases because the voltage drop across the armature resistance increases ($E = IR$).

In a series-wound generator, output voltage varies directly with load current. In the shunt-wound generator, output voltage varies inversely with load current. A combination of the two types can overcome the disadvantages of both. This combination of windings is called the compound-wound dc generator.

Compound-Wound Generators

Compound-wound generators have a series-field winding in addition to a shunt-field winding, as shown in figure 1-17. The shunt and series windings are wound on the same pole pieces.

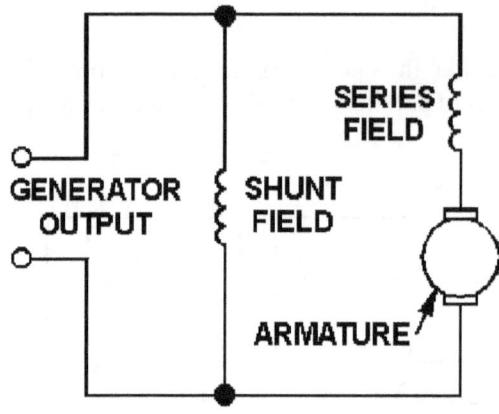

Figure 1-17.—Compound-wound generator.

In the compound-wound generator when load current increases, the armature voltage decreases just as in the shunt-wound generator. This causes the voltage applied to the shunt-field winding to decrease, which results in a decrease in the magnetic field. This same increase in load current, since it flows through the series winding, causes an increase in the magnetic field produced by that winding.

By proportioning the two fields so that the decrease in the shunt field is just compensated by the increase in the series field, the output voltage remains constant. This is shown in figure 1-18, which shows the voltage characteristics of the series-, shunt-, and compound-wound generators. As you can see, by proportioning the effects of the two fields (series and shunt), a compound-wound generator provides a constant output voltage under varying load conditions. Actual curves are seldom, if ever, as perfect as shown.

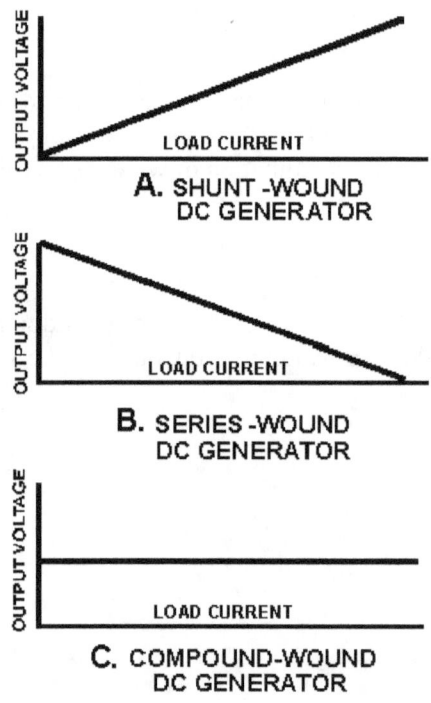

Figure 1-18.—Voltage output characteristics of the series-, shunt-, and compound-wound dc generators.

Q18. What are the three classifications of dc generators?

Q19. What is the main disadvantage of series generators?

GENERATOR CONSTRUCTION

Figure 1-19, views A through E, shows the component parts of dc generators. Figure 1-20 shows the entire generator with the component parts installed. The cutaway drawing helps you to see the physical relationship of the components to each other.

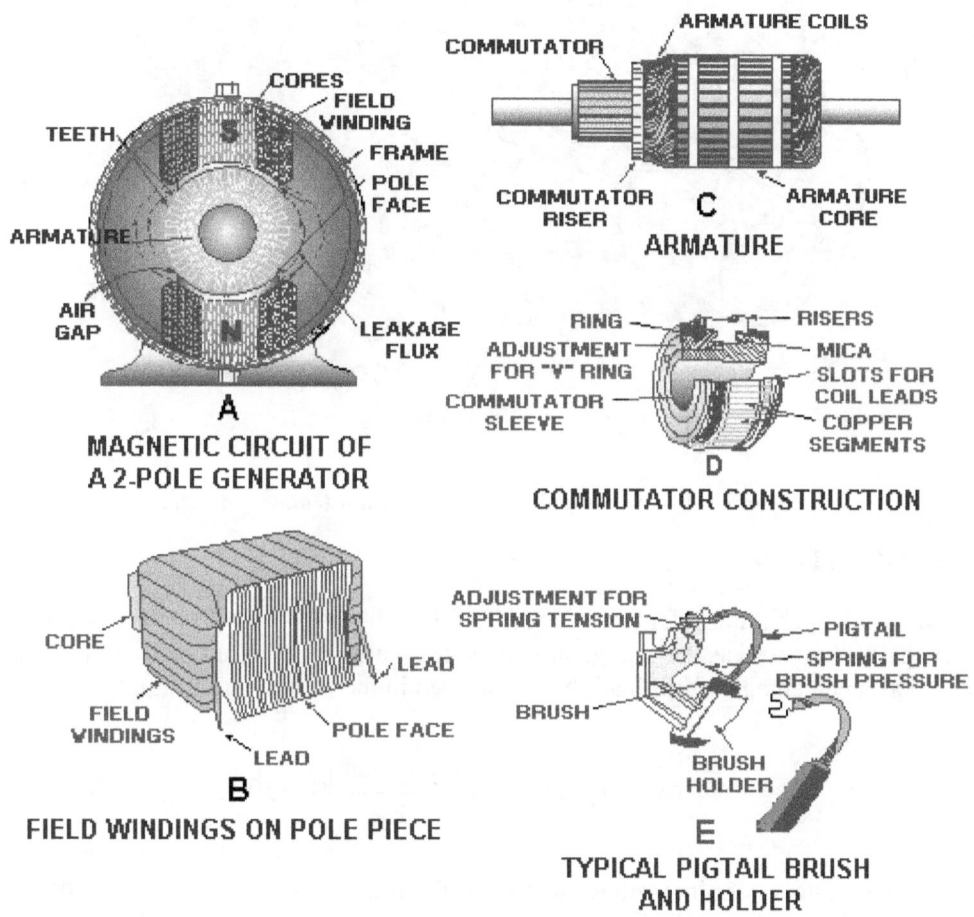

Figure 1-19.—Components of a dc generator.

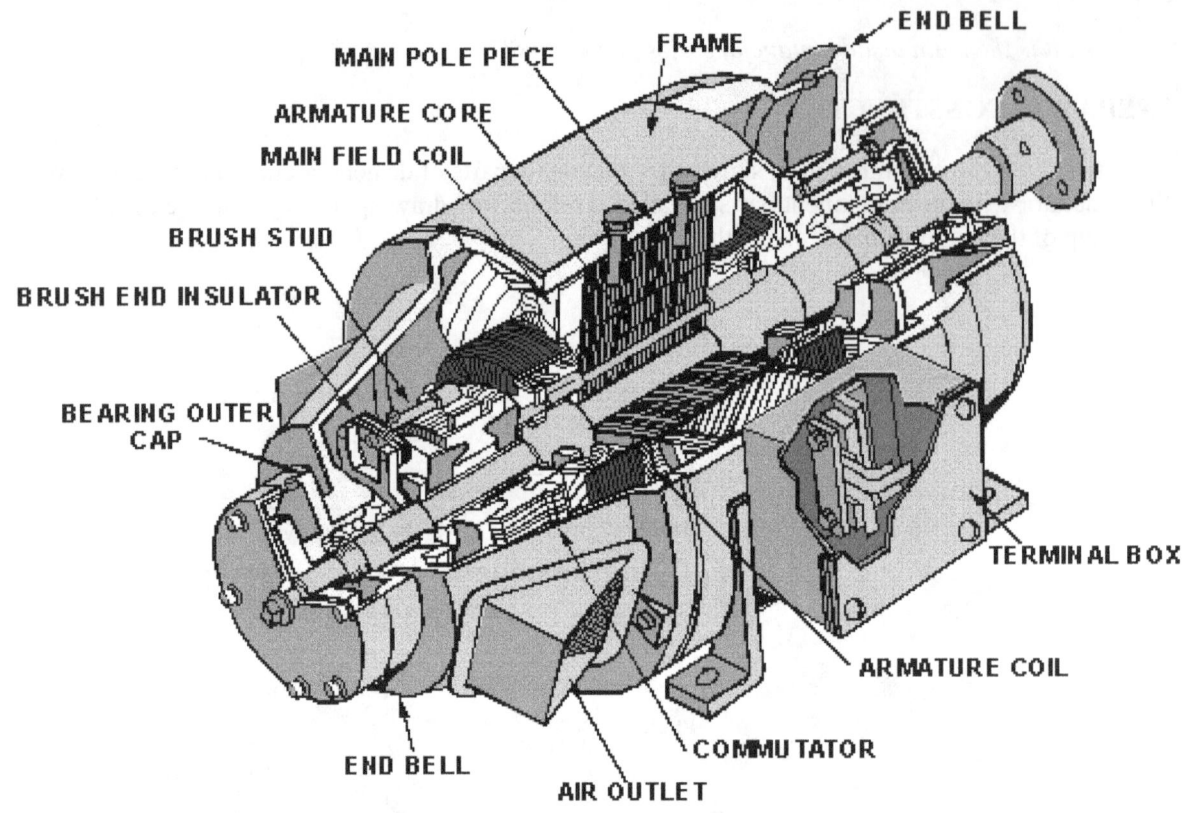

Figure 1-20.—Construction of a dc generator (cutaway drawing).

VOLTAGE REGULATION

The regulation of a generator refers to the VOLTAGE CHANGE that takes place when the load changes. It is usually expressed as the change in voltage from a no-load condition to a full-load condition, and is expressed as a percentage of full-load. It is expressed in the following formula:

$$\text{Percent of regulation} = \frac{(E_{nL} - E_{fL})}{E_{fL}} \times 100$$

where E_{nL} is the no-load terminal voltage and E_{fL} is the full-load terminal voltage of the generator. For example, to calculate the percent of regulation of a generator with a no-load voltage of 462 volts and a full-load voltage of 440 volts

Given:

- No-load voltage 462 V

- Full-load voltage 440 V

Solution:

$$\text{Percent of regulation} = \frac{(E_{nL} - E_{fL})}{E_{fL}} \times 100$$

$$\text{Percent of regulation} = \frac{(462V - 440V)}{440V} \times 100$$

$$\text{Percent of regulation} = \frac{22V}{440V} \times 100$$

$$\text{Percent of regulation} = .05 \times 100$$

$$\text{Regulation} = 5\%$$

NOTE: The lower the percent of regulation, the better the generator. In the above example, the 5% regulation represented a 22-volt change from no load to full load. A 1% change would represent a change of 4.4 volts, which, of course, would be better.

Q20. What term applies to the voltage variation from no-load to full-load conditions and is expressed as a percentage?

VOLTAGE CONTROL

Voltage control is either (1) manual or (2) automatic. In most cases the process involves changing the resistance of the field circuit. By changing the field circuit resistance, the field current is controlled. Controlling the field current permits control of the output voltage. The major difference between the various voltage control systems is merely the method by which the field circuit resistance and the current are controlled.

VOLTAGE REGULATION should not be confused with VOLTAGE CONTROL. As described previously, voltage regulation is an internal action occurring within the generator whenever the load changes. Voltage control is an imposed action, usually through an external adjustment, for the purpose of increasing or decreasing terminal voltage.

Manual Voltage Control

The hand-operated field rheostat, shown in figure 1-21, is a typical example of manual voltage control. The field rheostat is connected in series with the shunt field circuit. This provides the simplest method of controlling the terminal voltage of a dc generator.

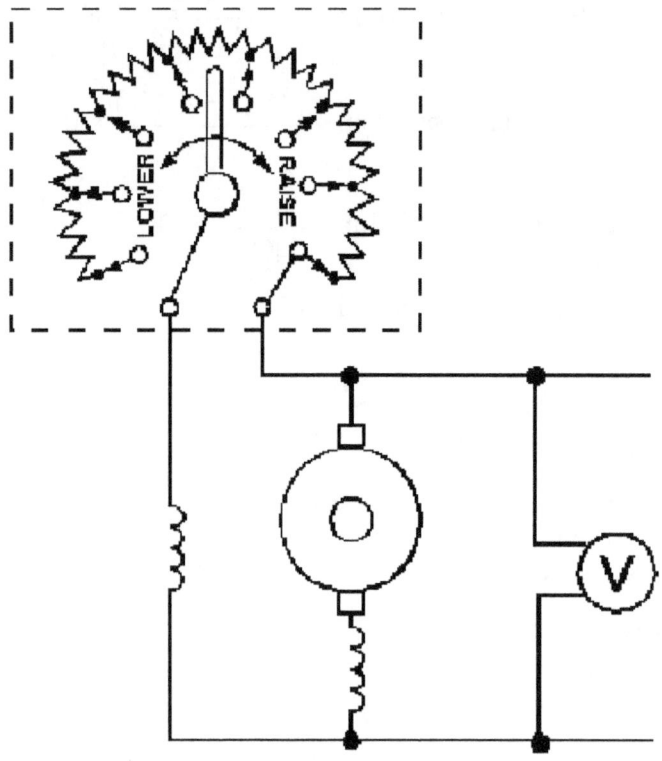

Figure 1-21.—Hand-operated field rheostat.

This type of field rheostat contains tapped resistors with leads to a multiterminal switch. The arm of the switch may be rotated to make contact with the various resistor taps. This varies the amount of resistance in the field circuit. Rotating the arm in the direction of the LOWER arrow (counterclockwise) increases the resistance and lowers the output voltage. Rotating the arm in the direction of the RAISE arrow (clockwise) decreases the resistance and increases the output voltage.

Most field rheostats for generators use resistors of alloy wire. They have a high specific resistance and a low temperature coefficient. These alloys include copper, nickel, manganese, and chromium. They are marked under trade names such as Nichrome, Advance, Manganin, and so forth. Some very large generators use cast-iron grids in place of rheostats, and motor-operated switching mechanisms to provide voltage control.

Automatic Voltage Control

Automatic voltage control may be used where load current variations exceed the built-in ability of the generator to regulate itself. An automatic voltage control device "senses" changes in output voltage and causes a change in field resistance to keep output voltage constant.

The actual circuitry involved in automatic voltage control will not be covered in this chapter. Whichever control method is used, the range over which voltage can be changed is a design characteristic of the generator. The voltage can be controlled only within the design limits.

PARALLEL OPERATION OF GENERATORS

When two or more generators are supplying a common load, they are said to be operating in parallel. The purpose of connecting generators in parallel is simply to provide more current than a single generator is capable of providing. The generators may be physically located quite a distance apart. However, they are connected to the common load through the power distribution system.

There are several reasons for operating generators in parallel. The number of generators used may be selected in accordance with the load demand. By operating each generator as nearly as possible to its rated capacity, maximum efficiency is achieved. A disabled or faulty generator may be taken off-line and replaced without interrupting normal operations.

Q21. What term applies to the use of two or more generators to supply a common load?

AMPLIDYNES

Amplidynes are special-purpose dc generators. They supply large dc currents, precisely controlled, to the large dc motors used to drive heavy physical loads, such as gun turrets and missile launchers.

The amplidyne is really a motor and a generator. It consists of a constant-speed ac motor (the prime mover) mechanically coupled to a dc generator, which is wired to function as a high-gain amplifier (an amplifier is a device in which a small input voltage can control a large current source). For instance, in a normal dc generator, a small dc voltage applied to the field windings is able to control the output of the generator. In a typical generator, a change in voltage from 0-volt dc to 3-volts dc applied to the field winding may cause the generator output to vary from 0-volt dc to 300-volts dc. If the 3 volts applied to the field winding is considered an input, and the 300 volts taken from the brushes is an output, there is a gain of 100. Gain is expressed as the ratio of output to input:

$$\text{Gain} = \frac{\text{output}}{\text{input}}$$

In this case 300 V • 3 V = 100. This means that the 3 volts output is 100 times larger than the input.

The following paragraphs explain how gain is achieved in a typical dc generator and how the modifications making the generator an amplidyne increase the gain to as high as 10,000.

The schematic diagram in figure 1-22 shows a separately excited dc generator. Because of the 10-volt controlling voltage, 10 amperes of current will flow through the 1-ohm field winding. This draws 100 watts of input power (P = IE).

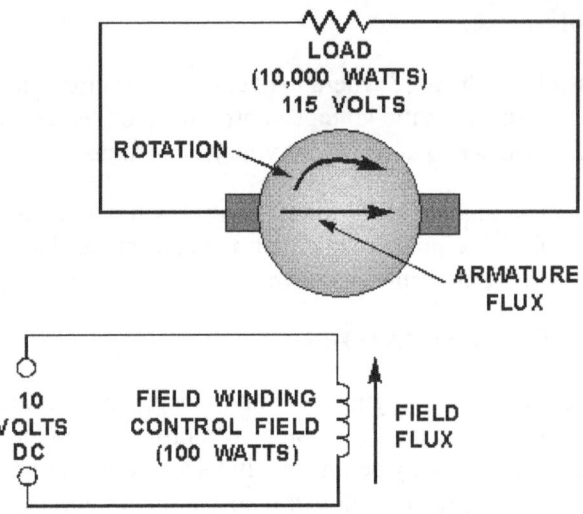

Figure 1-22.—Ordinary dc generator.

Assume that the characteristics of this generator enable it to produce approximately 87 amperes of armature current at 115 volts at the output terminals. This represents an output power of approximately 10,000 watts (P = IE). You can see that the power gain of this generator is 100. In effect, 100 watts controls 10,000 watts.

An amplidyne is a special type of dc generator. The following changes, for explanation purposes, will convert the typical dc generator above into an amplidyne.

The first step is to short the brushes together, as shown in figure 1-23. This removes nearly all of the resistance in the armature circuit.

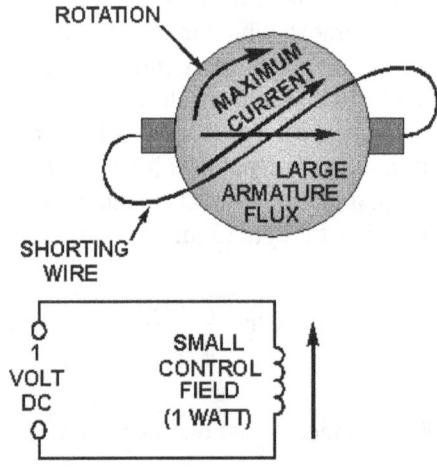

Figure 1-23.—Brushes shorted in a dc generator.

Because of the very low resistance in the armature circuit, a much lower control-field flux produces full-load armature current (full-load current in the armature is still about 87 amperes). The smaller control field now requires a control voltage of only 1 volt and an input power of 1 watt (1 volt across 1 ohm causes 1 ampere of current, which produces 1 watt of input power).

The next step is to add another set of brushes. These now become the output brushes of the amplidyne. They are placed against the commutator in a position perpendicular to the original brushes, as shown in figure 1-24. The previously shorted brushes are now called the "quadrature" brushes. This is because they are in quadrature (perpendicular) to the output brushes. The output brushes are in line with the armature flux. Therefore, they pick off the voltage induced in the armature windings at this point. The voltage at the output will be the same as in the original generator, 115 volts in our example.

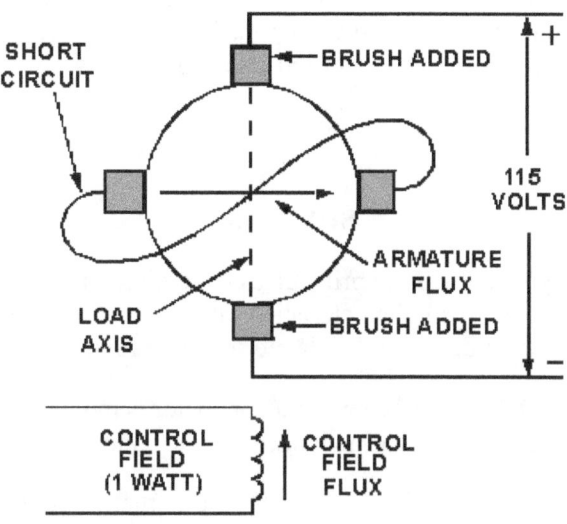

Figure 1-24.—Amplidyne load brushes.

As you have seen, the original generator produced a 10,000-watt output with a 100-watt input. The amplidyne produces the same 10,000-watt output with only a 1-watt input. This represents a gain of 10,000. The gain of the original generator has been greatly increased.

As previously stated, an amplidyne is used to provide large dc currents. The primary use of an amplidyne is in the positioning of heavy loads through the use of synchro/servo systems. Synchro/servo systems will be studied in a later module.

Assume that a very large turning force is required to rotate a heavy object, such as an antenna, to a very precise position. A low-power, relatively weak voltage representing the amount of antenna rotation required can be used to control the field winding of an amplidyne. Because of the amplidyne's ability to amplify, its output can be used to drive a powerful motor, which turns the heavy object (antenna). When the source of the input voltage senses the correct movement of the object, it drops the voltage to zero. The field is no longer strong enough to allow an output voltage to be developed, so the motor ceases to drive the object (antenna).

The above is an oversimplification and is not meant to describe a functioning system. The intent is to show a typical sequence of events between the demand for movement and the movement itself. It is meant to strengthen the idea that with the amplidyne, something large and heavy can be controlled very precisely by something very small, almost insignificant.

Q22. What is the purpose of a dc generator that has been modified to function as an amplidyne?

Q23. What is the formula used to determine the gain of an amplifying device?

Q24. What are the two inputs to an amplidyne?

SAFETY PRECAUTIONS

You must always observe safety precautions when working around electrical equipment to avoid injury to personnel and damage to equipment. Electrical equipment frequently has accessories that require separate sources of power. Lighting fixtures, heaters, externally powered temperature detectors, and alarm systems are examples of accessories whose terminals must be deenergized. When working on dc generators, you must check to ensure that all such circuits have been de-energized and tagged before you attempt any maintenance or repair work. You must also use the greatest care when working on or near the output terminals of dc generators.

SUMMARY

This chapter introduced you to the basic principles concerning direct current generators. The different types of dc generators and their characteristics were covered. The following information provides a summary of the major subjects of the chapter for your review.

MAGNETIC INDUCTION takes place when a conductor is moved in a magnetic field in such a way that it cuts flux lines, and a voltage (emf) is induced in the conductor.

THE LEFT-HAND RULE FOR GENERATORS states that when the thumb, forefinger, and middle finger of the left hand are extended at right angles to each other so that the thumb indicates the direction of movement of the conductor in the magnetic field, and the forefinger points in the direction of the flux lines (north to south), the middle finger shows the direction of induced emf in the conductor.

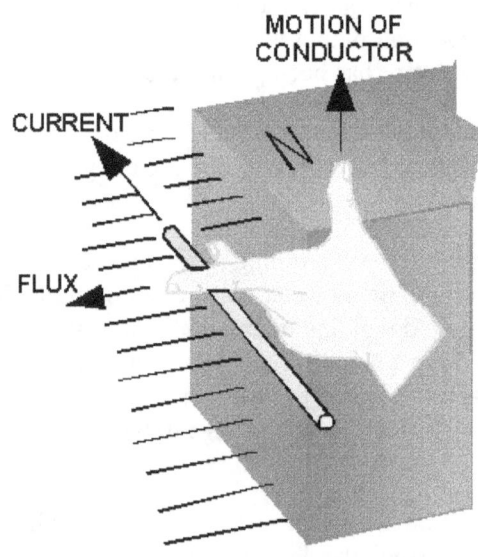

AN **ELEMENTARY GENERATOR** consists of a single coil rotated in a magnetic field. It produces an ac voltage.

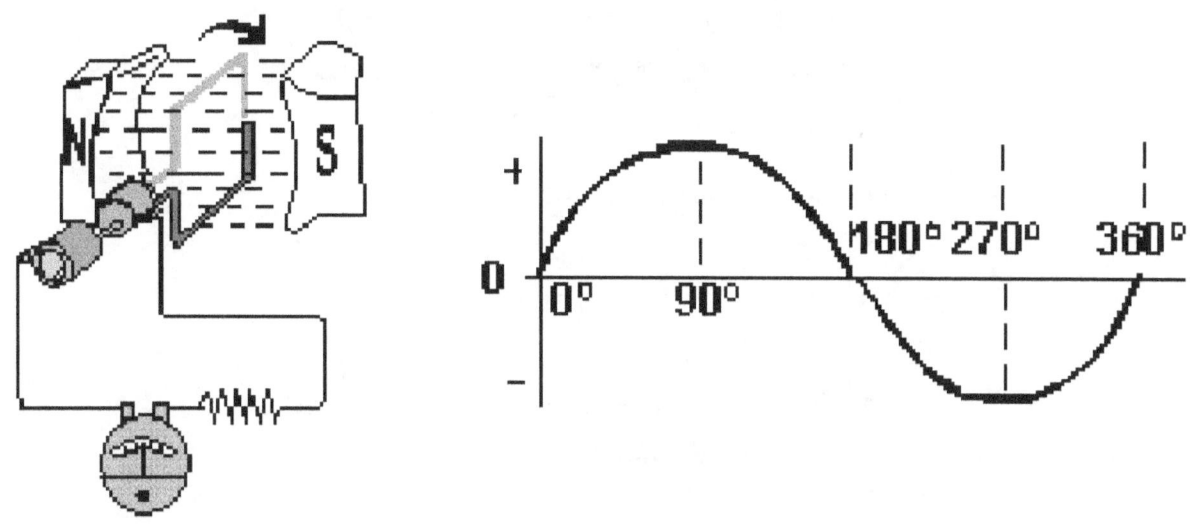

A **BASIC DC GENERATOR** results when you replace the slip rings of an elementary generator with a two-piece commutator, changing the output voltage to pulsating dc.

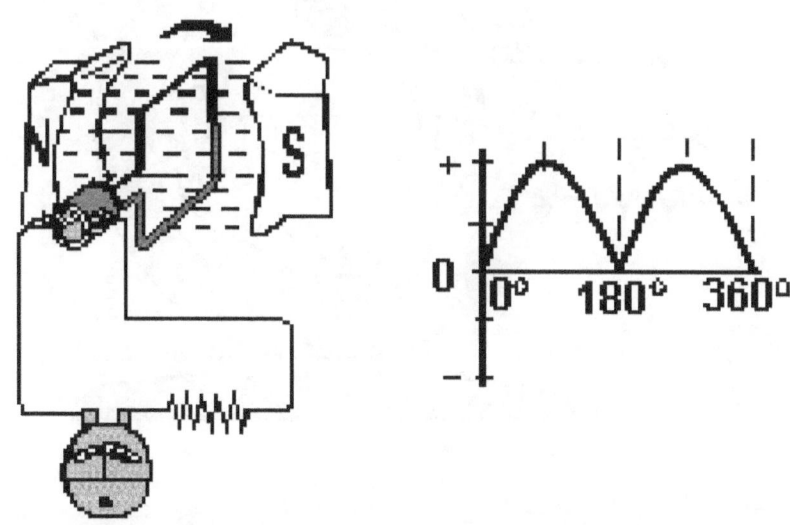

A **MULTIPLE COIL ARMATURE** (adding coils to the armature) decreases the ripple voltage in the output of a dc generator, and increases the output voltage.

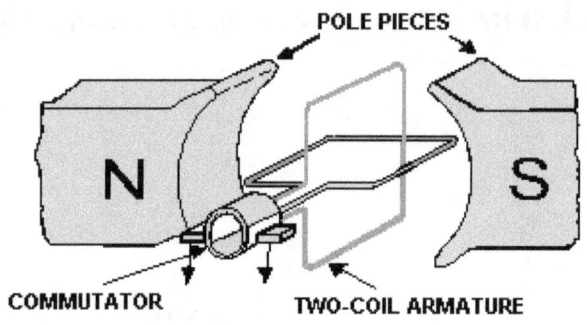

POLE PIECES

N S

COMMUTATOR   TWO-COIL ARMATURE

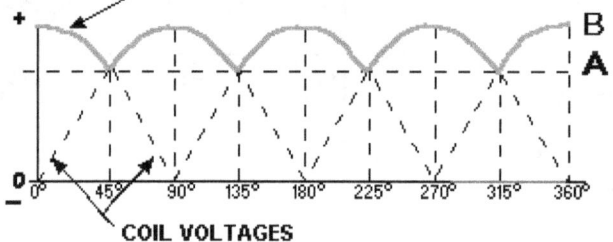

GENERATOR TERMINAL VOLTAGE

COIL VOLTAGES

A **MULTIPOLE GENERATOR** is the result of adding more field poles to a dc generator. They have much the same effect as adding coils to the armature. In practical generators, the poles are electromagnets.

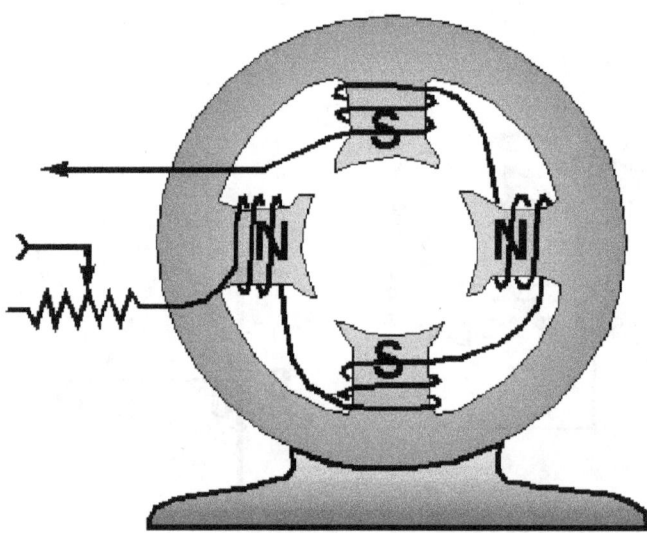

COMMUTATION is the process used to get direct current from a generator. The coil connections to the load must be reversed as the coil passes through the neutral plane. The brushes must be positioned so that commutation is accomplished without brush sparking.

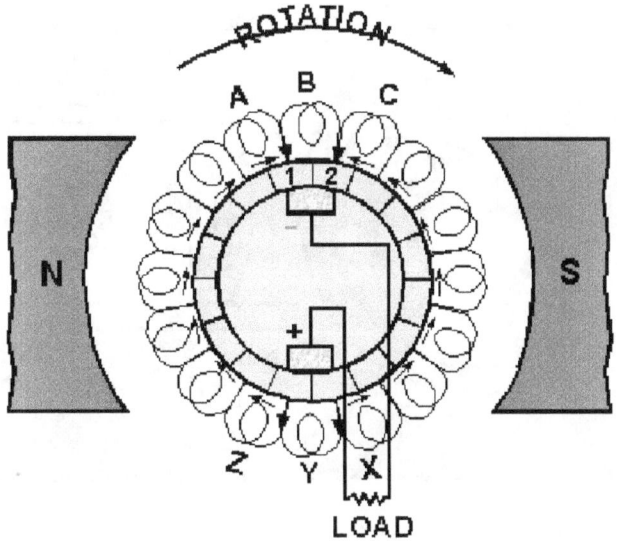

ARMATURE REACTION takes place when armature current causes the armature to become an electromagnet. The armature field disturbs the field from the pole pieces. This results in a shift of the neutral plane in the direction of rotation.

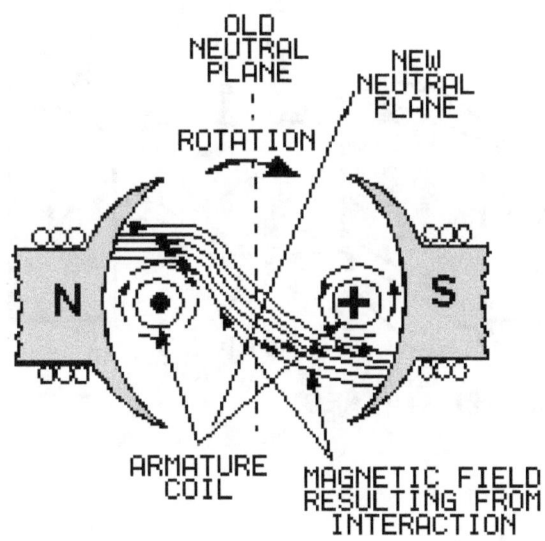

COMPENSATING WINDINGS AND INTERPOLES are used to counteract the effects of armature reaction. They are supplied by armature current and shift the neutral plane back to its original position.

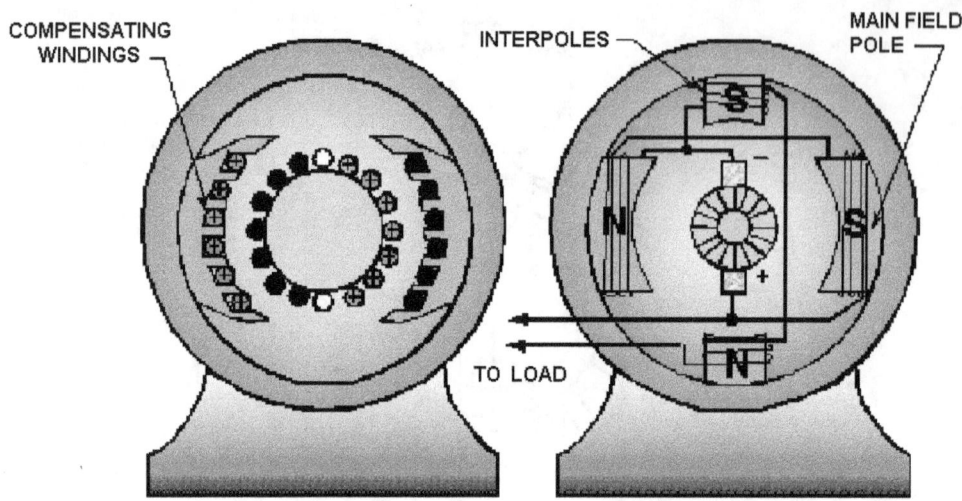

MOTOR REACTION is caused by the magnetic field that is set up in the armature. It tends to oppose the rotation of the armature, due to the attraction and repulsion forces between the armature field and the main field.

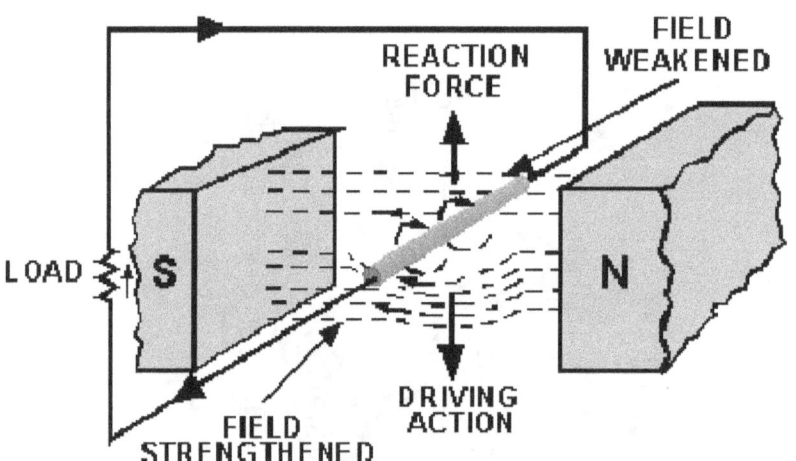

ARMATURE LOSSES in dc generator armatures affect the outputs. These losses are as follows:

1. Copper losses are simply I^2R (heat) losses caused by current flowing through the resistance of the armature windings.

2. Eddy currents are induced in core material and cause heat.

3. Hysteresis losses occur due to the rapidly changing magnetic fields in the armature, resulting in heat.

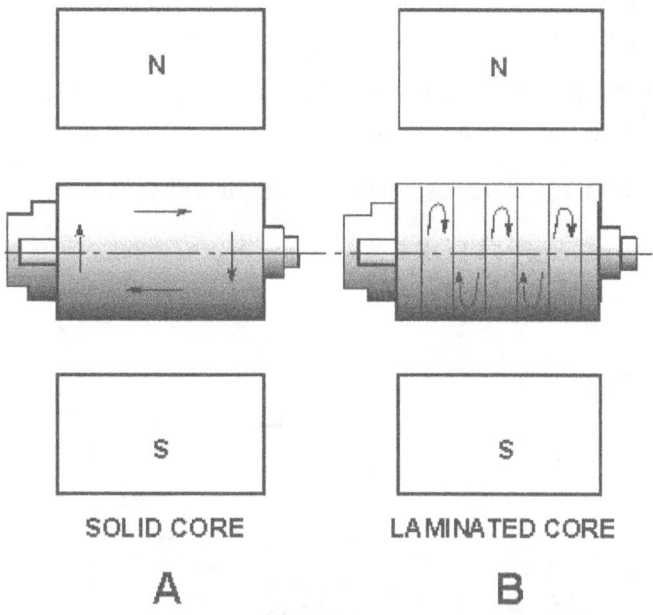

SOLID CORE   LAMINATED CORE

A   B

ARMATURE TYPES used in dc generators are the Gramme-ring (seldom used) and the drum-type, used in most applications.

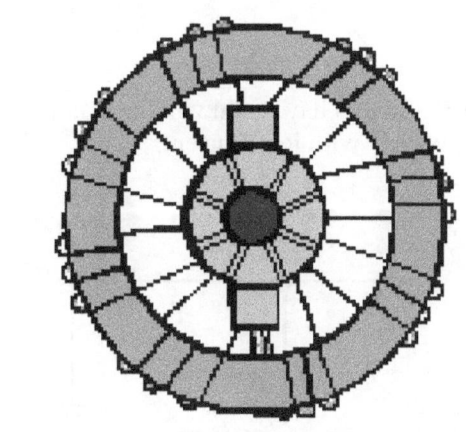

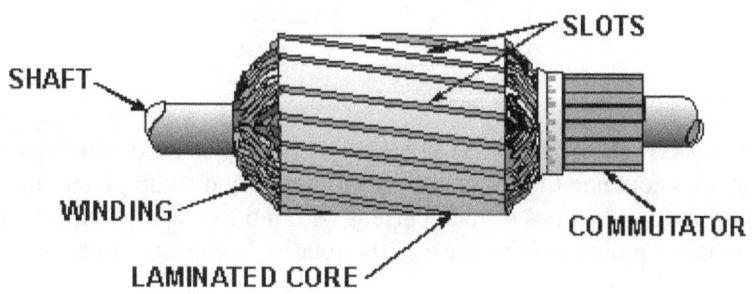

FIELD EXCITATION is the voltage applied to the main field windings. The current in the field coils determines the strength and the direction of the magnetic field.

SEPARATELY EXCITED GENERATORS receive current for field coils from an outside source such, as a battery or another dc generator.

SELF-EXCITED GENERATORS use their own output voltage to energize field coils.

SERIES-WOUND DC GENERATORS have field windings and armature windings connected in series. Outputs vary directly with load currents. Series-wound generators have few practical applications.

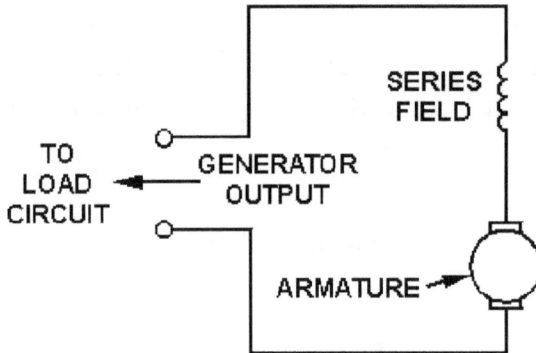

SHUNT-WOUND DC GENERATORS have field windings and armature windings connected in parallel (shunt). The output varies inversely with load current.

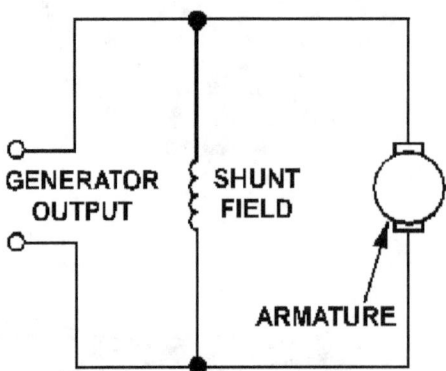

COMPOUND-WOUND DC GENERATORS have both series field windings and shunt field windings. These generators combine the characteristics of series and shunt generators. The output voltage remains relatively constant for all values of load current within the design of the generator. Compound generators are used in many applications because of the relatively constant voltage.

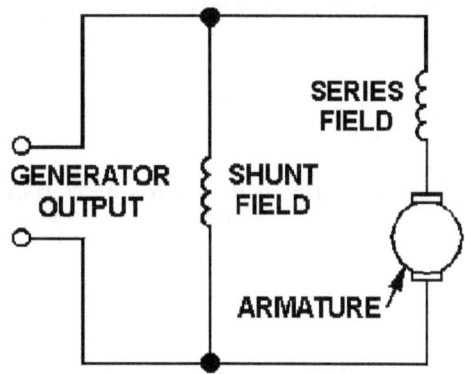

AMPLIDYNES are dc generators that are designed to act as high-gain amplifiers. By short-circuiting the brushes in a normal dc generator and adding another set of brushes perpendicular to the original ones, an amplidyne is formed. Its power output may be up to 10,000 times larger than the power input to its control windings.

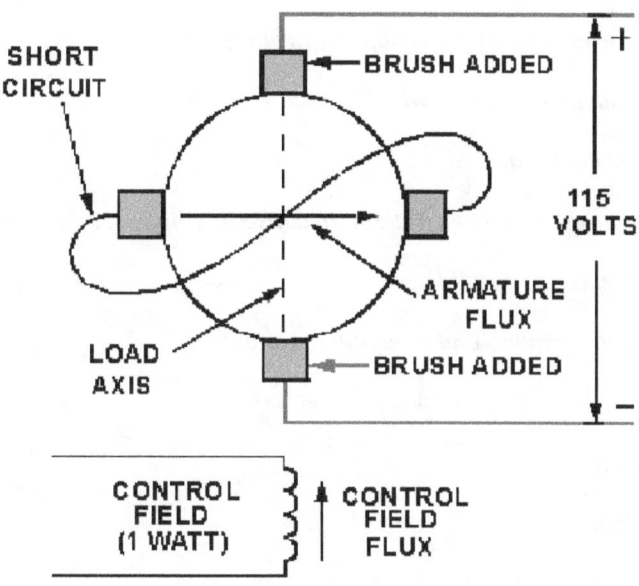

ANSWERS TO QUESTIONS Q1. THROUGH Q24.

A1.   Magnetic induction.

A2.   The left-hand rule for generators.

A3.   To conduct the currents induced in the armature to an external load.

A4.   No flux lines are cut.

A5.   A commutator

A6.   The point at which the voltage is zero across the two segments.

A7.   Two.

A8.   Four

A9.   By varying the input voltage to the field coils.

A10.   Improper commutation.

A11.   Distortion of the main field due to the effects of armature current.

A12.   To counter act armature reaction.

A13.   A force which causes opposition to applied turning force.

A14.   Resistance in the armature coils, which increases with temperature.

A15.   By laminating the core material.

A16.   Drum-type armatures are more efficient, because flux lines are cut by both sides of each coil.

A17.   Higher load currents are possible.

A18.   Series-wound, shunt-wound, and compound-wound.

A19.   Output voltage varies as the load varies.

A20.   Voltage regulation.

A21.   Parallel operation.

A22.   It can serve as a power amplifier.

A23.   Gain = output ÷ input.

A24.   The mechanical force applied to turn the amplidyne, and the electrical input signal.

CHAPTER 2

DIRECT CURRENT MOTORS

LEARNING OBJECTIVES

Upon completion of this chapter you will be able to:

1. State the factors that determine the direction of rotation in a dc motor.

2. State the right-hand rule for motors.

3. Describe the main differences and similarities between a dc generator and a dc motor.

4. Describe the cause and effect of counter emf in a dc motor.

5. Explain the term "load" as it pertains to an electric motor.

6. List the advantages and disadvantages of the different types of dc motors.

7. Compare the types of armatures and uses for each.

8. Discuss the means of controlling the speed and direction of a dc motor.

9. Describe the effect of armature reaction in a dc motor.

10. Explain the need for a starting resistor in a dc motor.

INTRODUCTION

The dc motor is a mechanical workhorse, that can be used in many different ways. Many large pieces of equipment depend on a dc motor for their power to move. The speed and direction of rotation of a dc motor are easily controlled. This makes it especially useful for operating equipment, such as winches, cranes, and missile launchers, which must move in different directions and at varying speeds.

PRINCIPLES OF OPERATION

The operation of a dc motor is based on the following principle:

A current-carrying conductor placed in a magnetic field, perpendicular to the lines of flux, tends to move in a direction perpendicular to the magnetic lines of flux.

There is a definite relationship between the direction of the magnetic field, the direction of current in the conductor, and the direction in which the conductor tends to move. This relationship is best explained by using the RIGHT-HAND RULE FOR MOTORS (fig. 2-1).

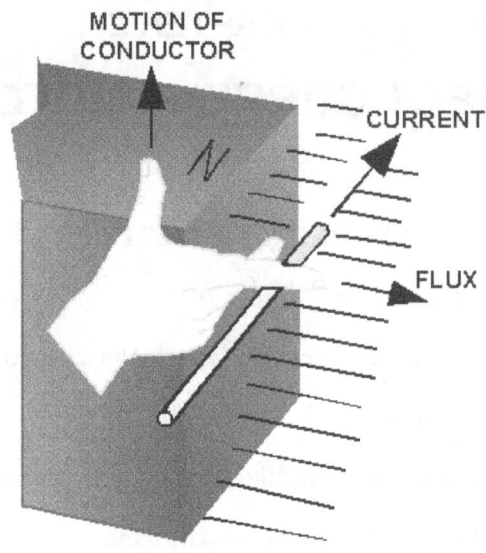

Figure 2-1.—Right-hand rule for motors.

To find the direction of motion of a conductor, extend the thumb, forefinger, and middle finger of your right hand so they are at right angles to each other. If the forefinger is pointed in the direction of magnetic flux (north to south) and the middle finger is pointed in the direction of current flow in the conductor, the thumb will point in the direction the conductor will move.

Stated very simply, a dc motor rotates as a result of two magnetic fields interacting with each other. The armature of a dc motor acts like an electromagnet when current flows through its coils. Since the armature is located within the magnetic field of the field poles, these two magnetic fields interact. Like magnetic poles repel each other, and unlike magnetic poles attract each other. As in the dc generator, the dc motor has field poles that are stationary and an armature that turns on bearings in the space between the field poles. The armature of a dc motor has windings on it just like the armature of a dc generator. These windings are also connected to commutator segments. A dc motor consists of the same components as a dc generator. In fact, most dc generators can be made to act as motors, and vice versa.

Look at the simple dc motor shown in figure 2-2. It has two field poles, one a north pole and one a south pole. The magnetic lines of force extend across the opening between the poles from north to south.

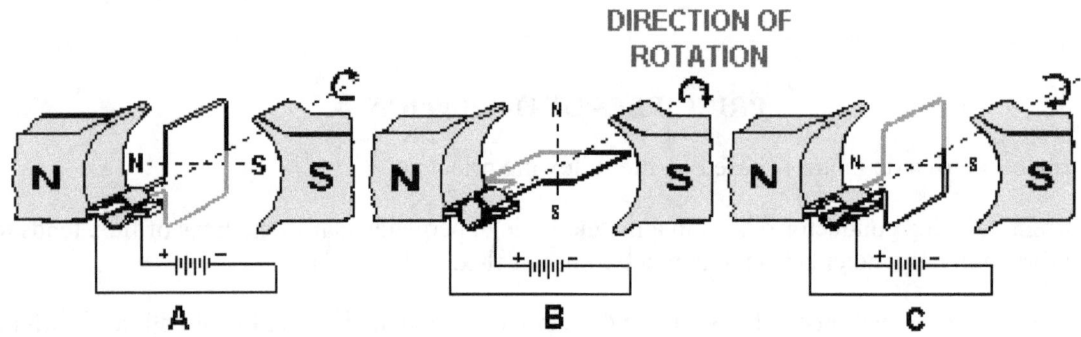

Figure 2-2.—Dc motor armature rotation.

The armature in this simple dc motor is a single loop of wire, just as in the simple armature you studied at the beginning of the chapter on dc generators. The loop of wire in the dc motor, however, has current flowing through it from an external source. This current causes a magnetic field to be produced. This field is indicated by the dotted line through the loops. The loop (armature) field is both attracted and repelled by the field from the field poles. Since the current through the loop goes around in the direction of the arrows, the north pole of the armature is at the upper left, and the south pole of the armature is at the lower right, as shown in figure 2-2, (view A). Of course, as the loop (armature) turns, these magnetic poles turn with it. Now, as shown in the illustrations, the north armature pole is repelled from the north field pole and attracted to the right by the south field pole. Likewise, the south armature pole is repelled from the south field pole and is attracted to the left by the north field pole. This action causes the armature to turn in a clockwise direction, as shown in figure 2-2 (view B).

After the loop has turned far enough so that its north pole is exactly opposite the south field pole, the brushes advance to the next segments. This changes the direction of current flow through the armature loop. Also, it changes the polarity of the armature field, as shown in figure 2-2 (view C). The magnetic fields again repel and attract each other, and the armature continues to turn.

In this simple motor, the momentum of the rotating armature carries the armature past the position where the unlike poles are exactly lined up. However, if these fields are exactly lined up when the armature current is turned on, there is no momentum to start the armature moving. In this case, the motor would not rotate. It would be necessary to give a motor like this a spin to start it. This disadvantage does not exist when there are more turns on the armature, because there is more than one armature field. No two armature fields could be exactly aligned with the field from the field poles at the same time.

Q1. What factors determine the direction of rotation in a dc motor?

Q2. The right-hand rule for motors is used to find the relationship between what motor characteristics?

Q3. What are the differences between the components of a dc generator and a dc motor?

COUNTER EMF

While a dc motor is running, it acts somewhat like a dc generator. There is a magnetic field from the field poles, and a loop of wire is turning and cutting this magnetic field. For the moment, disregard the fact that there is current flowing through the loop of wire from the battery. As the loop sides cut the magnetic field, a voltage is induced in them, the same as it was in the loop sides of the dc generator. This induced voltage causes current to flow in the loop.

Now, consider the relative direction between this current and the current that causes the motor to run. First, check the direction the current flows as a result of the generator action taking place (view A of fig. 2-2). (Apply the left-hand rule for generators which was discussed in the last chapter.) Using the left hand, hold it so that the forefinger points in the direction of the magnetic field (north to south) and the thumb points in the direction that the black side of the armature moves (up). Your middle finger then points out of the paper (toward you), showing the direction of current flow caused by the generator action in the black half of the armature. This is in the direction opposite to that of the battery current. Since this generator-action voltage is opposite that of the battery, it is called "counter emf." (The letters emf stand for electromotive force, which is another name for voltage.) The two currents are flowing in opposite directions. This proves that the battery voltage and the counter emf are opposite in polarity.

At the beginning of this discussion, we disregarded armature current while explaining how counter emf was generated. Then, we showed that normal armature current flowed opposite to the current created by the counter emf. We talked about two opposite currents that flow at the same time. However, this is a bit oversimplified, as you may already suspect. Actually, only one current flows. Because the counter emf can never become as large as the applied voltage, and because they are of opposite polarity as we have seen, the counter emf effectively cancels part of the armature voltage. The single current that flows is armature current, but it is greatly reduced because of the counter emf.

In a dc motor, there is always a counter emf developed. This counter emf cannot be equal to or greater than the applied battery voltage; if it were, the motor would not run. The counter emf is always a little less. However, the counter emf opposes the applied voltage enough to keep the armature current from the battery to a fairly low value. If there were no such thing as counter emf, much more current would flow through the armature, and the motor would run much faster. However, there is no way to avoid the counter emf.

Q4. *What causes counter emf in a dc motor?*

Q5. *What motor characteristic is affected by counter emf?*

MOTOR LOADS

Motors are used to turn mechanical devices, such as water pumps, grinding wheels, fan blades, and circular saws. For example, when a motor is turning a water pump, the water pump is the load. The water pump is the mechanical device that the motor must move. This is the definition of a motor load.

As with electrical loads, the mechanical load connected to a dc motor affects many electrical quantities. Such things as the power drawn from the line, amount of current, speed, efficiency, etc., are all partially controlled by the size of the load. The physical and electrical characteristics of the motor must be matched to the requirements of the load if the work is to be done without the possibility of damage to either the load or the motor.

Q6. *What is the load on a dc motor?*

PRACTICAL DC MOTORS

As you have seen, dc motors are electrically identical to dc generators. In fact, the same dc machine may be driven mechanically to generate a voltage, or it may be driven electrically to move a mechanical load. While this is not normally done, it does point out the similarities between the two machines. These similarities will be used in the remainder of this chapter to introduce you to practical dc motors. You will immediately recognize series, shunt, and compound types of motors as being directly related to their generator counterparts.

SERIES DC MOTOR

In a series dc motor, the field is connected in series with the armature. The field is wound with a few turns of large wire, because it must carry full armature current. The circuit for a series dc motor is shown in figure 2-3.

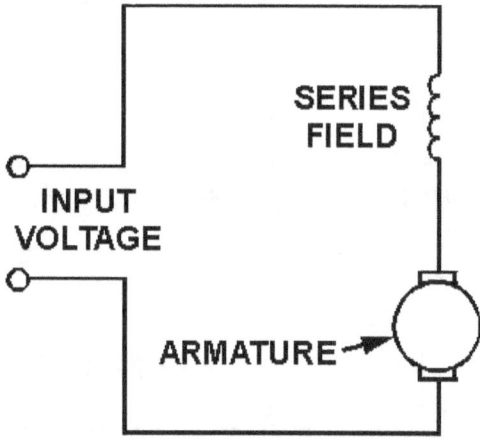

Figure 2-3.—Series-wound dc motor.

This type of motor develops a very large amount of turning force, called torque, from a standstill. Because of this characteristic, the series dc motor can be used to operate small electric appliances, portable electric tools, cranes, winches, hoists, and the like.

Another characteristic is that the speed varies widely between no-load and full-load. Series motors cannot be used where a relatively constant speed is required under conditions of varying load.

A major disadvantage of the series motor is related to the speed characteristic mentioned in the last paragraph. The speed of a series motor with no load connected to it increases to the point where the motor may become damaged. Usually, either the bearings are damaged or the windings fly out of the slots in the armature. There is a danger to both equipment and personnel. Some load must ALWAYS be connected to a series motor before you turn it on. This precaution is primarily for large motors. Small motors, such as those used in electric hand drills, have enough internal friction to load themselves.

A final advantage of series motors is that they can be operated by using either an ac or dc power source. This will be covered in the chapter on ac motors.

Q7. What is the main disadvantage of a series motor?

Q8. What is the main advantage of a series motor?

SHUNT MOTOR

A shunt motor is connected in the same way as a shunt generator. The field windings are connected in parallel (shunt) with the armature windings. The circuit for a shunt motor is shown in figure 2-4.

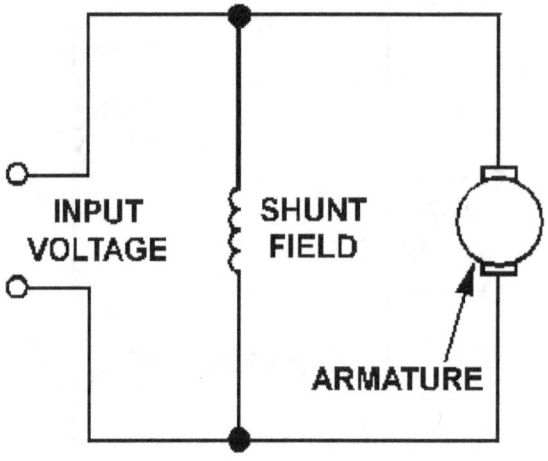

Figure 2-4.—Shunt-wound dc motor.

Once you adjust the speed of a dc shunt motor, the speed remains relatively constant even under changing load conditions. One reason for this is that the field flux remains constant. A constant voltage across the field makes the field independent of variations in the armature circuit.

If the load on the motor is increased, the motor tends to slow down. When this happens, the counter emf generated in the armature decreases. This causes a corresponding decrease in the opposition to battery current flow through the armature. Armature current increases, causing the motor to speed up. The conditions that established the original speed are reestablished, and the original speed is maintained.

Conversely, if the motor load is decreased, the motor tends to increase speed; counter emf increases, armature current decreases, and the speed decreases.

In each case, all of this happens so rapidly that any actual change in speed is slight. There is instantaneous tendency to change rather than a large fluctuation in speed.

Q9. What advantage does a shunt motor have over a series motor?

COMPOUND MOTOR

A compound motor has two field windings, as shown in figure 2-5. One is a shunt field connected in parallel with the armature; the other is a series field that is connected in series with the armature. The shunt field gives this type of motor the constant speed advantage of a regular shunt motor. The series field gives it the advantage of being able to develop a large torque when the motor is started under a heavy load. It should not be a surprise that the compound motor has both shunt- and series-motor characteristics.

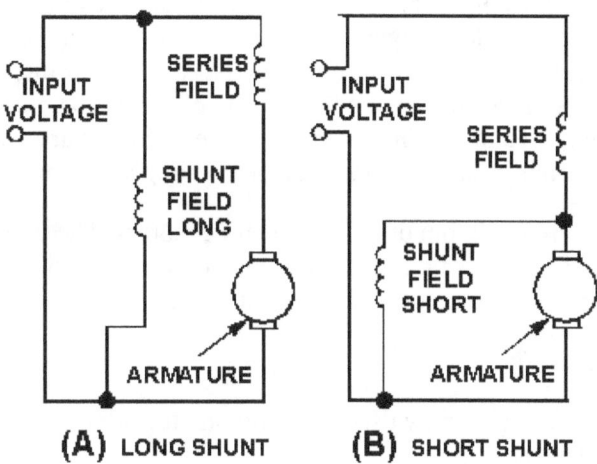

Figure 2-5.—Compound-wound dc motor.

When the shunt field is connected in parallel with the series field and armature, it is called a "long shunt" as shown in figure 2-5, (view A). Otherwise, it is called a "short shunt", as shown in figure 2-5, (view B).

TYPES OF ARMATURES

As with dc generators, dc motors can be constructed using one of two types of armatures. A brief review of the Gramme-ring and drum-wound armatures is necessary to emphasize the similarities between dc generators and dc motors.

GRAMME-RING ARMATURE

The Gramme-ring armature is constructed by winding an insulated wire around a soft-iron ring (fig. 2-6). Eight equally spaced connections are made to the winding. Each of these is connected to a commutator segment. The brushes touch only the top and bottom segments. There are two parallel paths for current to follow — one up the left side and one up the right side. These paths are completed through the top brush back to the positive lead of the battery.

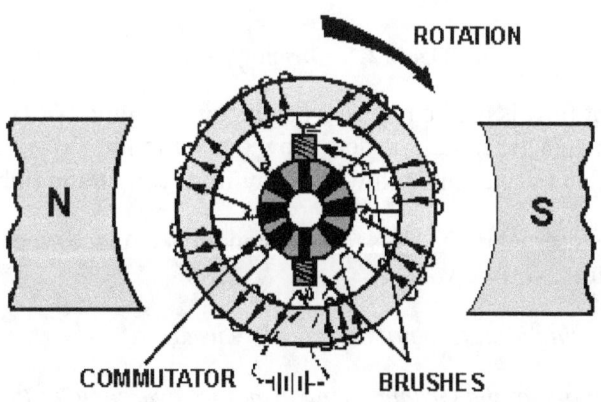

Figure 2-6.—Gramme-ring armature.

To check the direction of rotation of this armature, you should use the right-hand rule for motors. Hold your thumb, forefinger, and middle finger at right angles. Point your forefinger in the direction of field flux; in this case, from left to right. Now turn your wrist so that your middle finger points in the direction that the current flows in the winding on the outside of the ring. Note that current flows into the page (away from you) in the left-hand windings and out of the page (toward you) in the right-hand windings. Your thumb now points in the direction that the winding will move.

The Gramme-ring armature is seldom used in modem dc motors. The windings on the inside of the ring are shielded from magnetic flux, which causes this type of armature to be inefficient. The Gramme-ring armature is discussed primarily to help you better understand the drum-wound armature.

DRUM-WOUND ARMATURE

The drum-wound armature is generally used in ac motors. It is identical to the drum winding discussed in the chapter on dc generators.

If the drum-wound armature were cut in half, an end view at the cut would resemble the drawing in figure 2-7, (view A), Figure 2-7, (view B) is a side view of the armature and pole pieces. Notice that the length of each conductor is positioned parallel to the faces of the pole pieces. Therefore, each conductor of the armature can cut the maximum flux of the motor field. The inefficiency of the Gramme-ring armature is overcome by this positioning.

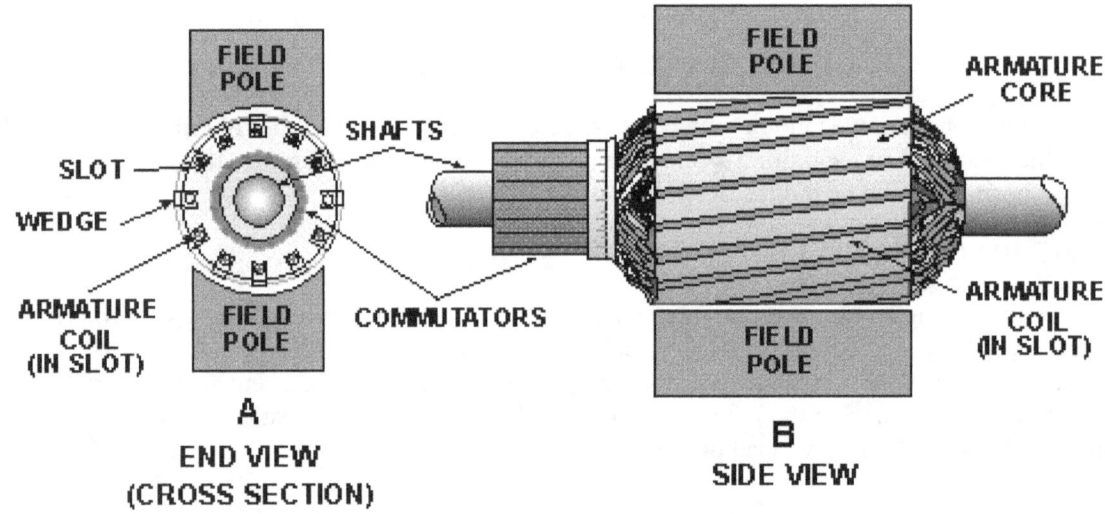

Figure 2-7.—Drum-type armature.

The direction of current flow is marked in each conductor in figure 2-7, (view A) as though the armature were turning in a magnetic field. The dots show that current is flowing toward you on the left side, and the crosses show that the current is flowing away from you on the right side.

Strips of insulation are inserted in the slots to keep windings in place when the armature spins. These are shown as wedges in figure 2-7, (view A).

Q10. Why is the Gramme-ring armature not more widely used?

Q11. How is the disadvantage of the Gramme-ring armature overcome in the drum-wound armature?

DIRECTION OF ROTATION

The direction of rotation of a dc motor depends on the direction of the magnetic field and the direction of current flow in the armature. If either the direction of the field or the direction of current flow through the armature is reversed, the rotation of the motor will reverse. However, if both of these factors are reversed at the same time, the motor will continue rotating in the same direction. In actual practice, the field excitation voltage is reversed in order to reverse motor direction.

Ordinarily, a motor is set up to do a particular job that requires a fixed direction of rotation. However, there are times when it is necessary to change the direction of rotation, such as a drive motor for a gun turret or missile launcher. Each of these must be able to move in both directions. Remember, the connections of either the armature or the field must be reversed, but not both. In such applications, the proper connections are brought out to a reversing switch.

Q12.   In a dc motor that must be able to rotate in both directions, how is the direction changed?

MOTOR SPEED

A motor whose speed can be controlled is called a variable-speed motor; dc motors are variable-speed motors. The speed of a dc motor is changed by changing the current in the field or by changing the current in the armature.

When the field current is decreased, the field flux is reduced, and the counter emf decreases. This permits more armature current. Therefore, the motor speeds up. When the field current is increased, the field flux is increased. More counter emf is developed, which opposes the armature current. The armature current then decreases, and the motor slows down.

When the voltage applied to the armature is decreased, the armature current is decreased, and the motor again slows down. When the armature voltage and current are both increased, the motor speeds up.

In a shunt motor, speed is usually controlled by a rheostat connected in series with the field windings, as shown in figure 2-8. When the resistance of the rheostat is increased, the current through the field winding is decreased. The decreased flux momentarily decreases the counter emf. The motor then speeds up, and the increase in counter emf keeps the armature current the same. In a similar manner, a decrease in rheostat resistance increases the current flow through the field windings and causes the motor to slow down.

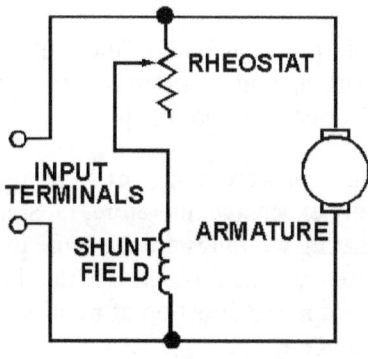

Figure 2-8.—Controlling motor speed.

In a series motor, the rheostat speed control may be connected either in parallel or in series with the armature windings. In either case, moving the rheostat in a direction that lowers the voltage across the armature lowers the current through the armature and slows the motor. Moving the rheostat in a direction that increases the voltage and current through the armature increases motor speed.

Q13. What is the effect on motor speed if the field current is increased?

ARMATURE REACTION

You will remember that the subject of armature reaction was covered in the previous chapter on dc generators. The reasons for armature reaction and the methods of compensating for its effects are basically the same for dc motors as for dc generators.

Figure 2-9 reiterates for you the distorting effect that the armature field has on the flux between the pole pieces. Notice, however, that the effect has shifted the neutral plane backward, against the direction of rotation. This is different from a dc generator, where the neutral plane shifted forward in the direction of rotation.

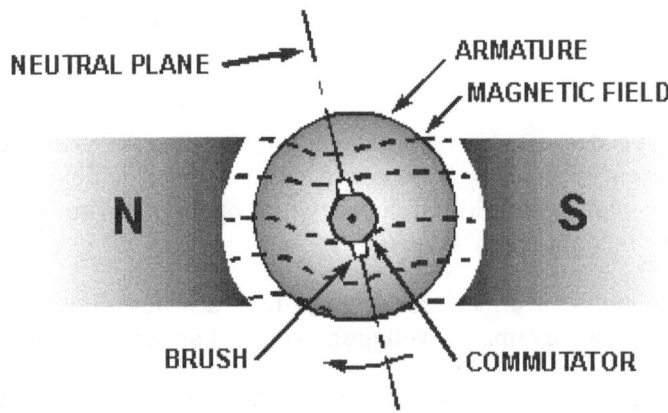

Figure 2-9.—Armature reaction.

As before, the brushes must be shifted to the new neutral plane. As shown in figure 2-9, the shift is counterclockwise. Again, the proper location is reached when there is no sparking from the brushes.

Q14. Armature reaction in a dc motor causes a shift of the neutral plane in which direction?

Compensating windings and interpoles, two more "old" subjects, cancel armature reaction in dc motors. Shifting brushes reduces sparking, but it also makes the field less effective. Canceling armature reaction eliminates the need to shift brushes in the first place.

Compensating windings and interpoles are as important in motors as they are in generators. Compensating windings are relatively expensive; therefore, most large dc motors depend on interpoles to correct armature reaction. Compensating windings are the same in motors as they are in generators. Interpoles, however, are slightly different. The difference is that in a generator the interpole has the same polarity as the main pole AHEAD of it in the direction of rotation. In a motor the interpole has the same polarity as the main pole FOLLOWING it.

The interpole coil in a motor is connected to carry the armature current the same as in a generator. As the load varies, the interpole flux varies, and commutation is automatically corrected as the load changes. It is not necessary to shift the brushes when there is an increase or decrease in load. The brushes are located on the no-load neutral plane. They remain in that position for all conditions of load.

Q15. What current flows in the interpole windings?

The dc motor is reversed by reversing the direction of the current in the armature. When the armature current is reversed, the current through the interpole is also reversed. Therefore, the interpole still has the proper polarity to provide automatic commutation.

MANUAL AND AUTOMATIC STARTERS

Because the dc resistance of most motor armatures is low (0.05 to 0.5 ohm), and because the counter emf does not exist until the armature begins to turn, it is necessary to use an external starting resistance in series with the armature of a dc motor to keep the initial armature current to a safe value. As the armature begins to turn, counter emf increases; and, since the counter emf opposes the applied voltage, the armature current is reduced. The external resistance in series with the armature is decreased or eliminated as the motor comes up to normal speed and full voltage is applied across the armature.

Controlling the starting resistance in a dc motor is accomplished either manually, by an operator, or by any of several automatic devices. The automatic devices are usually just switches controlled by motor speed sensors. Automatic starters are not covered in detail in this module.

Q16. What is the purpose of starting resistors?

SUMMARY

This chapter presented the operating principles and characteristics of direct-current motors. The following information provides a summary of the main subjects for review.

The main **PRINCIPLE OF A DC MOTOR** is that current flow through the armature coil causes the armature to act as a magnet. The armature poles are attracted to field poles of opposite polarity, causing the armature to rotate.

The **CONSTRUCTION** of a dc motor is almost identical to that of a dc generator, both physically and electrically. In fact, most dc generators can be made to act as dc motors, and vice versa.

COMMUTATION IN A DC MOTOR is the process of reversing armature current at the moment when unlike poles of the armature and field are facing each other, thereby reversing the polarity of the armature field. Like poles of the armature and field then repel each other, causing armature rotation to continue.

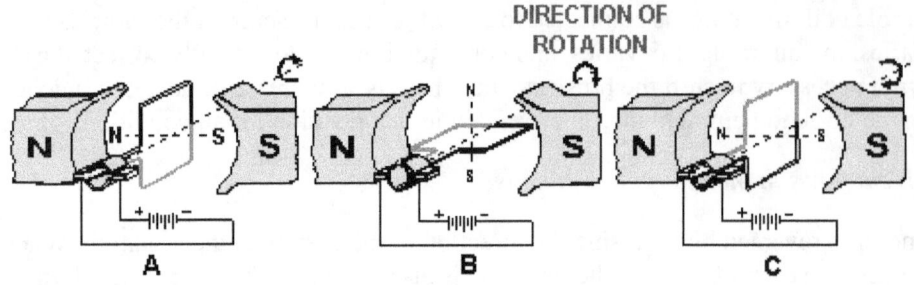

A   B   C

COUNTER-ELECTROMOTIVE FORCE is generated in a dc motor as armature coils cut the field flux. This emf opposes the applied voltage, and limits the flow of armature current.

In **SERIES MOTORS**, the field windings are connected in series with the armature coil. The field strength varies with changes in armature current. When its speed is reduced by a load, the series motor develops greater torque. Its starting torque is greater than other types of dc motors. Its speed varies widely between full-load and no-load. Unloaded operation of large machines is dangerous.

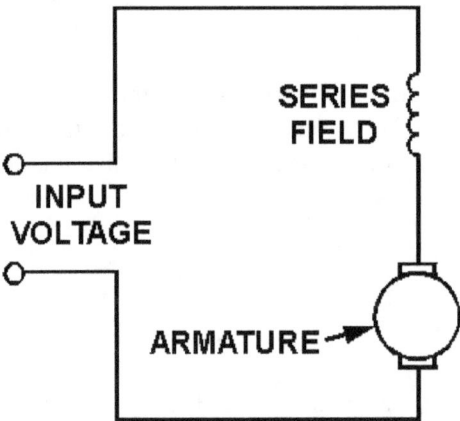

In **SHUNT MOTORS**, the field windings are connected in parallel (shunt) across the armature coil. The field strength is independent of the armature current. Shunt-motor speed varies only slightly with changes in load, and the starting torque is less than that of other types of dc motors.

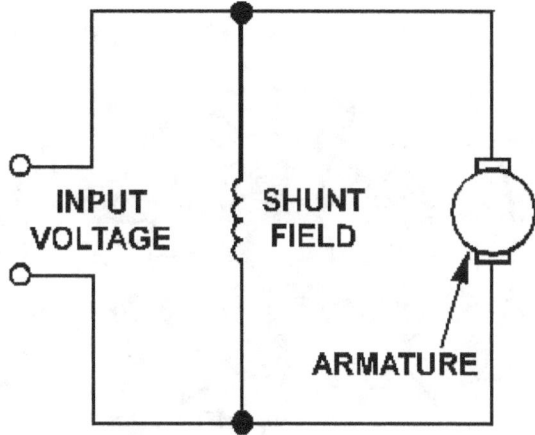

In **COMPOUND MOTORS**, one set of field windings is connected in series with the armature, and one set is connected in parallel. The speed and torque characteristics are a combination of the desirable characteristics of both series and shunt motors.

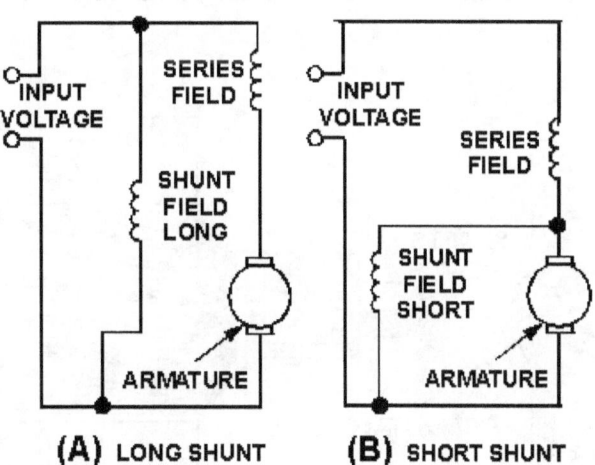

LOAD on a motor is the physical object to be moved by the motor.

DC MOTOR ARMATURES are of two types. They are the Gramme-ring and the drum-wound types.

THE GRAMME-RING ARMATURE is inefficient since part of each armature coil is prevented from cutting flux lines. Gramme-ring wound armatures are seldom used for this reason.

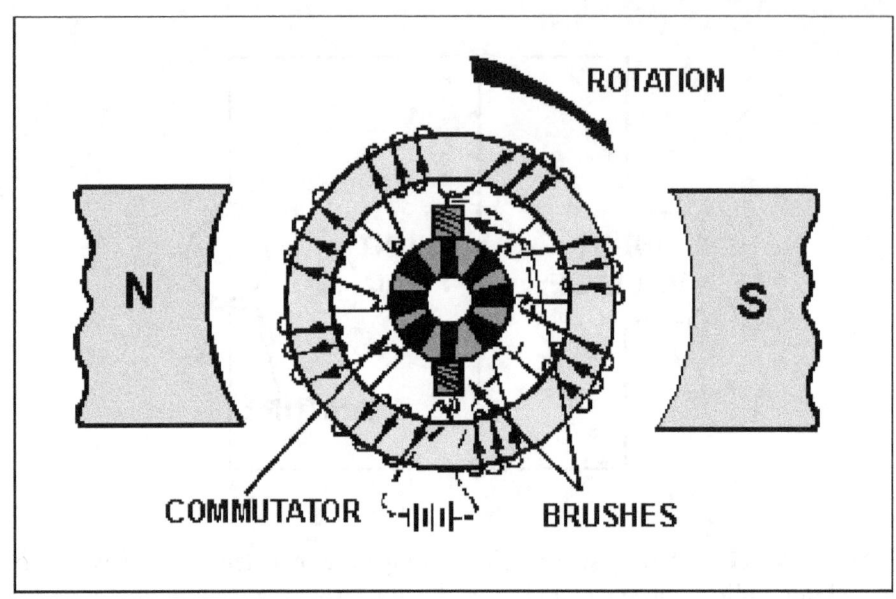

THE DRUM-WOUND ARMATURE consists of coils actually wound around the armature core so that all coil surfaces are exposed to the magnetic field. Nearly all dc motors have drum-wound armatures.

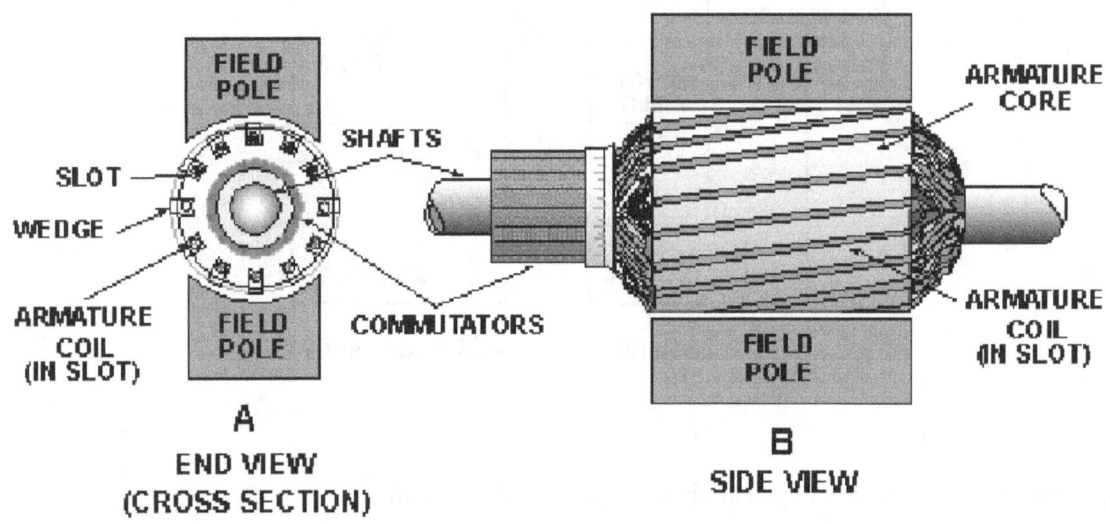

MOTOR REVERSAL in a dc motor can be accomplished by reversing the field connections or by reversing the armature connections. If both are reversed, rotation will continue in the original direction.

SPEED CONTROL IN A DC MOTOR is maintained by varying the resistance either in series with the field coil or in series with the armature coil. Increasing shunt-field circuit resistance increases motor speed. Increasing the armature circuit resistance decreases motor speed.

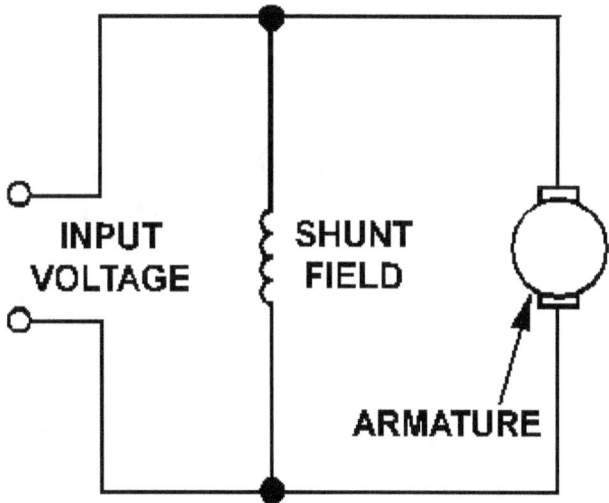

ARMATURE REACTION is the distortion of the main field in a motor by the armature field. This causes the neutral plane to be shifted in the direction opposite to that of armature rotation. Interpoles and compensating windings are used to reduce the effect of armature reaction on motor operation.

STARTING RESISTORS are necessary since the dc resistance of a motor armature is very low. Excessive current will flow when dc voltage is first applied unless current is limited in some way. Adding resistance in series with the armature windings reduces initial current. It may then be removed after counter emf has been built up.

ANSWERS TO QUESTIONS Q1. THROUGH Q16.

A1. *Direction of armature current, and direction of magnetic flux in field.*

A2. *Direction of conductor movement (rotation), direction of flux, and the direction of current flow.*

A3. *There are no differences.*

A4. *Generator action.*

A5. *Speed.*

A6. *The device to be driven by the motor.*

A7. *It must have a load connected to avoid damage from excess speed.*

A8. *High torque (turning force) at low speed.*

A9. *It maintains a constant speed under varying loads.*

A10. *Only outside of coils cut flux (inefficient).*

A11. *By winding the armature in a way that places the entire coil where it is exposed to maximum flux.*

A12. *By reversing either field or armature connections.*

A13.   *Motor will slow down.*

A14.   *Opposite the rotation.*

A15.   *Armature current.*

A16.   *To limit armature current until counter emf builds up.*

www.ingramcontent.com/pod-product-compliance
Lightning Source LLC
Chambersburg PA
CBHW080613180526
45168CB00007B/2902

* 9 7 8 1 5 0 8 4 9 7 0 2 8 *